Water Pricing and Public-Private Partnership

There is no question that water pricing and public-private partnership can improve water management practices in the future. However, it is not the panacea as the proponents argue, nor the disaster as its opponents forecast. *Water Pricing and Public-Private Partnership* provides a comprehensive and objective assessment as to what works, where, why and under what conditions, as well as what does not work and why. It also assesses the social, economic, equity and institutional implications.

The book provides an in-depth analysis and assessment of the main issues and constraints on pricing of water and participation of the private sector and their affect on water supply in North and South America and Western Europe. With case studies from Argentina, Brazil, the USA and West European countries, among many others, we gain a comprehensive and objective account of the economic and social consequences.

This collection transcends the current dogmatic debate on these complex issues resulting in a cohesive set of carefully selected essays. This vital book is the result of The Third World Centre for Water Management and the Inter-American Development Bank's decision to objectively and critically assess the experiences in these areas from North and South America and Western Europe.

This is a special issue of the journal *International Journal of Water Resources Development*, which was entitled *Water Pricing and Public-Private Partnership in the Americas.*

Asit K. Biswas is President of the Third World Centre for Water Management, Mexico and Editor-in-Chief of the *International Journal of Water Resources Development.*

Cecilia Tortajada is Vice-President of the Third World Centre for Water Management, Mexico and Editor of the *International Journal of Water Resources Development.*

WATER PRICING AND PUBLIC-PRIVATE PARTNERSHIP

Edited by
Asit K. Biswas and Cecilia Tortajada

LONDON AND NEW YORK

First published 2005 by Routledge
2 Park Square, Milton Park, Abingdon, Oxon, OX14 4RN

Simultaneously published in the USA and Canada
by Routledge
270 Madison Ave, New York, NY 10016

Routledge is an imprint of the Taylor & Francis Group

Transferred to Digital Printing 2007

© 2005 Asit K. Biswas, Cecilia Tortajada, Benedito P.F. Braga
and Deigo Rodriguez

Typeset in Times 10/12pt in Europe by Alden Group, Oxford

British Library Cataloguing in Publication Data
A catalogue record for this book is available from the British Library

Library of Congress Cataloging in Publication Data

ISBN10: 0–415–37121–X (hbk)
ISBN10: 0–415–43673–7 (pbk)

ISBN13: 978–0–415–37121–6 (hbk)
ISBN13: 978–0–415–43673–1 (pbk)

Printed and bound by CPI Antony Rowe, Eastbourne

CONTENTS

Introduction

ASIT K. BISWAS

Introduction

Leonardo da Vinci, the eminent renaissance scholar and philosopher, said, "water is the driver of nature". Many people in the past have considered it to be an over-statement, but at the beginning of the third millennium, it is difficult to disagree with Leonardo's views. During the past five years, water has been increasingly considered to be one of the most critical natural resources issues for the early part of the 21st century. Analyses of all the current water-related trends indicate that the overall global water situation, at least for the next decade, if not for longer, is likely to deteriorate even further.

Strangely, however, neither the water and the development professionals, nor the international development-related institutions, realized or appreciated the seriousness of the global water situation as late as 1995, even though a few observers have been predicting this possibility for at least two decades. This can probably be best illustrated by the fact that the seriousness of the water crisis was not even a minor issue for discussion at either the International Conference on Water and the Environment that was organized by the United Nations System in Dublin, or at the United Nations Conference on Environment and Development at Rio de Janeiro. Held during the first half of 1992, both are considered to be landmark events of the 1990s.

It is now being increasingly realized that the Dublin Conference was poorly planned and organized, with very little appreciation or understanding of the potential water problems of the world, let alone finding and recommending solutions that could be implemented. Furthermore, as the Dublin Conference was expected to make the necessary inputs to the Rio discussions, not surprisingly, water did not receive adequate consideration at Rio.

For all practical purposes, at Rio, water was basically ignored by all the Heads of States, whose primary interests were focused on issues like climate change, biodiversity and deforestation. Water was at best a very minor issue during the plenary discussions. The chapter on water in Agenda 21, one of the main outputs at Rio, is not only the longest but is also the most poorly formulated. Thus, despite the rhetoric of the institutions and the individuals associated with the organization of the Dublin and the Rio Conferences, their impacts on global, regional or local water management have not been discernible. In all probability, development in the water sector would not have been materially different even if these two events had not occurred.

Water Crisis

In retrospect, it appears that the global water crisis was brought to prominence in the international agenda, not through the work of international organizations such as the United Nations agencies, but primarily through the leadership and efforts of the Stockholm Water Symposium, an annual water event that is attended by leading water professionals from all over the world. By regularly focusing on the importance and relevance of water to the future socio-economic development of the world and its importance to environmental conservation, the Stockholm Water Symposium managed to convince the water profession that the global water situation, at least during the early part of the 21st century is likely to get worse, and thus must be considered seriously.

By the late 1990s, water became an important issue for discussion in the international political agenda. By the beginning of the 21st century, many international organizations suddenly decided that the world is heading for a water crisis that will be unprecedented in human history. Accordingly, water received much higher emphasis at the Johannesburg Summit of 2002, compared to at the Rio Conference. The Johannesburg Summit was organized by the United Nations to review the progress made since the Rio Conference a decade ago. The Committee on Sustainable Development of the United Nations decided to give high priority to water. The UN General Assembly declared the year 2003 as the 'Year of the Freshwater'. These decisions should have been taken at least a decade ago.

Regrettably, just as in the early 1990s water did not receive adequate attention, the issue of the water crisis is somewhat overblown at present.

While predicting the future is an extremely hazardous business, one item can be predicted with complete certainty: the world in 2025 will be vastly different compared to what it is today. Among the main driving forces that are likely to contribute to these changes are rapidly changing demographic conditions, concurrent urbanization and ruralization in the developing world, technological advances in water and non-water related areas, information and communication revolution, speed and extent of globalization, improvements in human capital and national and international policies. While some of the pathways through which the water sector will be affected by these driving forces are known, or can be predicted, many of the pathways are unknown at present.

While it is not realized at present, it is highly likely that water development and management practices will change more during the next 20 years, compared to the changes that have been witnessed during the past 2000 years (Biswas, 1999). So far, the water profession has generally ignored the global forces outside the water sector that are already shaping the future availability of resources and its use and management practices. These impacts are likely to increase very significantly during the next quarter of a century, and yet the water profession has not yet started to factor in these new developments, and their impacts on the global water scene, even in a cursory manner. Much of the impetus for these changes will come from outside the water sector, and the water profession will have to adjust rapidly to these changes. All these developments are likely to revolutionize future water planning and management policies and practices. Because of these changes, some of which now can be foreseen, it is possible to become cautiously optimistic of the global water future.

Water Pricing

One of the issues that is likely to have major impacts on water use and management practices in the future is water pricing, a topic that is considered to be somewhat controversial at present. The World Commission on Water for the 21st Century in its report (2000) said categorically:

> Commission members agreed that the single most immediate and important measure that we can recommend is the systematic adoption of full-cost pricing for water services.

The reasons for this recommendation were outlined by the Commission as follows:

- Far too few public resources are devoted to public goods, especially environmental enhancement;
- Free water leads to wasted water;
- Considerable resources invested in the water and sanitation sectors, estimated at $30 billion a year, are used inefficiently;
- Governments in developing counties cannot meet the investment demands for water services now, let alone for the future, yet they are under-investing in public goods.

The Commission then went on to say:

> The situation is clear, without full-cost pricing the present vicious cycle of waste, inefficiency, and lack of services for the poor will continue. There will be little investment from the private sector, services will be of poor quality and rationed, and there will be little left for investing in water quality and other environmental improvements. The corollary is that there could be a 'virtuous cycle', too. This could be one in which users pay for the services that want, in which the urban utilities and irrigation agencies provide these services efficiently and accountably, in which users pay the costs of these services, in which investors place their money, and in which public funds are used primarily for public purposes.

Some four years after this report was issued, an objective assessment indicates that the Commission (the author was a member of this Commission) basically got the thrust of its arguments right, even though water pricing was one of the two issues that attracted maximum debate during the Second and Third World Water Forums at The Hague, in March 2000, and Kyoto, in March 2003. The second most controversial issue of this report was the recommendation of private sector involvement in the water sector, which is also a focus of this issue of the journal. Of course, this is not surprising since in many ways these two issues are interlinked. The private sector would have no interest in managing a water concession unless they could recover their investment costs, and also make a reasonable return on the capital invested. Without water pricing, there will be no private sector interest in managing large urban water concessions.

It is now being widely accepted that water services can no longer be free to all the consumers, or could be provided to them ad infinitum and universally on a highly subsidized basis. The era when water could be provided free is now mostly over. Efficient water management is simply not possible unless demands are managed properly, consistently and equitably. One of the tools for efficient demand management is unquestionably water pricing.

This, of course, does not mean that we have all the answers on how water should be priced for different consumers and for different uses. How can we ensure that the poor have adequate access to reliable water and sanitation services and affordable prices, but concurrently the rich are not subsidized? How, by whom, and through what processes should these services be managed so as to ensure the objectives of provision of reliable services, economic efficiency and maximizing social welfare are met concurrently? These and many other associated questions need to be answered satisfactorily, based on observed facts and scientific analyses, and not on dogmatic views. What is becoming increasingly clear is that there are no universal solutions. Each case needs to be analysed and judged on its merits and constraints. What may be working well for England and Wales, where the entire water services were sold to the private sector, may not be an appropriate solution for Argentina or Mexico. Equally, within a country there are likely to be several alternatives, especially if the countries are large and diverse in terms of climatic, economic and social conditions, like USA, Brazil, China or India. Furthermore, solutions may also vary with time in the same location, depending upon the prevailing socio-political conditions and economic considerations.

Public–Private Partnership

The World Commission on Water also made several comments on the possible roles of the private sector in managing water resources in the future. The report pointed out that:

- Adequate incentives should be provided to the private sector to contribute where it is well equipped to do so;
- The private sector will bring neither its money nor its management skills and know-how, unless it can operate in a predictable, transparent regulatory environment, and unless it can get a reasonable return on its investments, without undue political interference;
- The private sector can bring in additional financing to the water sector, which the public sector alone cannot provide;
- The Private sector can considerably improve the current poor technical and financial performances of most water utilities in the developing world.

Both at the Second World Water Forum in the Hague in 2000, and the Third World Water Forum in Japan in 2003, a main contention of the activists, who strongly opposed both water pricing and private sector participation, was that these two issues are interrelated. The point was made repeatedly that water pricing is a 'code word' for handing over an essential public service to the private sector, which will then make unseemly profits at the cost of the poor.

During the Hague Forum, it was widely assumed that a few multinational corporations (4–6) would 'control' the water services sectors of urban areas of the world. They would become so big and powerful that the public regulators would not be able to control them. By the Third World Water Forum, a scant three years after the Hague event, the goal-pasts had shifted. Whereas in 2000 a few multinational companies were increasing their outreach at a very rapid pace, by 2003, the same companies were beating a retreat from the developing world. Saddled by huge debts and significant losses in many concessions, and facing a steeply declining share process, most multinational water companies had decided to significantly curtail their ambitious plans of the recent past. The post-2000 period also witnessed very significant changes in these companies, through mergers, demergers and bankruptcies.

Since the World Commission on Water issued its report some three years ago, the knowledge-base in this area has improved significantly. Based on the latest analyses, the following can be observed:

- The private sector currently serves about 4–7% (estimates vary) of urban water consumers in the developing world. This percentage is likely to increase in the coming years, but most likely at a much slower rate than what was anticipated even only two years ago.
- Under all foreseeable conditions, the vast majority of domestic urban consumers will continue to receive their water and wastewater services from the publicly-run water companies, at least for the next 15 years.

Because of controversies that are common in this area, the Third World Centre for Water Management decided to carry out a series of in-depth analyses on the performances of the private sector in various developing countries. Based on these analyses, following conclusions can be drawn.

First, there are many forms of private sector involvement. These could range from outright sales of assets to the private sector, as was the case for England and Wales, to provision of management concessions to run water supply and wastewater collection and treatment facilities over a fixed number of years (current concessions range from 3 to 60 years), to outsourcing of specific activities. Since England and Wales sold outright all its assets to the private sector in 1989, no other country has followed this model. Some 15 years after this privatization in England and Wales, there is currently no agreement amongst water professionals as to its actual impacts on the consumers and quality of services provided. The assessments available at present range from highly favourable to equally highly deplorable.

In contrast to outright sale of assets, the use of management contracts to the private sector for specific period of years has proliferated during the past 5–7 years. Here again no universal judgement can be made. Some concessions have been very successful, but equally others have been dismal failures. Results have sometimes varied even within a single country (for example, in Morocco, Casablanca could be considered to be a success but not Rabat), and in one instance at least in the same metropolitan area (half of Manila works, but the other half does not).

There could also be a time dimension to the effectiveness of the private sector involvement. Thus, Buenos Aires was a good example of private sector involvement, until the economic meltdown in Argentina changed all the boundary conditions. Recently, the concessionaire had to write off nearly $500 million of its investments in this city.

Thus, there is no single model of private sector participation which could be appropriate to all cases in one country, let alone for the entire world.

Second, after rapid expansion in the award of concessions to manage water supply and wastewater systems to the private sector in recent years, the rate of award of new similar arrangements has slowed down very perceptibly since the year 2002. Near terms prospects for new concessions do not look encouraging. On 9 January 2003, one of the major multinational groups announced a 5-point action plan for 2003–2004, which included:

- reduction of debt mainly by selling assets;
- cost reduction;
- new investments to be financed by cash flow, which means that its new annual investments will fall from €8 billion to €4 billion;
- reorganization; and
- reducing its exposure in developing countries by one-third.

Third, competitive pressure from the multinational water companies has improved the performances of the public sector companies in many developed as well as developing countries. This is an advantage that the water professionals have mostly missed. It is expected that the performances of many public sector water companies would improve quite considerably in the future, certainly at a much higher rate than has been the norm for the past 20 years. Without the threat of the private sector competition, it is highly unlikely that the performances of the public sector companies would have improved in such a remarkable fashion in recent years, and over such a short period of time.

Finally, performances of public and private sector companies should not be generalized. By most criteria, the best water utility in the world continues to be Singapore, a public sector endeavour. Even within the private sector companies, their performances have varied very significantly, ranging from excellent to very dismal. Equally, the performances of individual multinational water companies have varied from one city to another, and also could vary over time even in the same city. Thus, any objective analysis of the current state of affairs will have to conclude that the performances of the public sector are not necessarily uniformly bad, and equally the achievements of all the private sector companies are not necessarily uniformly good. Each case should be judged on its merits and demerits, and over a specific period of time. Thus, both the high priests of the private sector who claim that the private sector will solve all the problems, and the die-hard social activists who claim that private sector has no role to play in water supply and wastewater management are wrong. Each project must be judged by its performances, which must be based on objective analyses of facts, and must not make generalized statements based on dogmas and/or hidden agendas.

Concluding Remarks

The papers for this issue were specially commissioned by the Third World Centre for Water Management to review the status of the use of water pricing in different parts of the world, and also what have been the global experiences in public–private partnership in the water sector. Supported by the Inter-American Development Bank and Agencia Nacional de Aguas of the Government of Brazil, the only objective of the analyses in this special issue is not to promote water pricing and public–private partnerships as panaceas, but to discuss what are their comparative advantages and disadvantages compared to the traditional practices of the recent years. Based on such information and experiences, individuals in specific locations can then make informed decisions as to what could be the most optimal decisions for the case in question.

References

Biswas, A. K. (1999) Water crisis: current perceptions and future realities, *Water International*, 24(4), pp. 363–367.
World Commission on Water for the 21st Century (2000) *A Water Secure World: Vision for Water, Life and the Environment* (Marseilles: World Water Council).

Water Pricing: An Outsider's Perspective

LUIS E. GARCÍA
Fairfax, VA, USA

Introduction

The human perception about water has been gradually changing during the last three decades. It could be said that a change in paradigms can be traced in the declarations and resolutions of many international and regional conferences and meetings. Among the more relevant are perhaps the United Nation's 1977 Mar del Plata Water Conference Action Plan, the 1992 Water and Environment Dublin Conference, the 1992 Rio de Janeiro's Agenda 21, the 1998 Paris Water and Sustainable Development declaration, The declaration of The Hague World Water Forum of 2000, the 2001 Bonn declaration, the Millennium Development Goals, the Johannesburg Earth Summit declaration and the 2003 Japan World Water Forum declaration.

The international consensus developed after Dublin,[1] as to what should be done for a more efficient and equitable use of the water resources, has become a widely known revered benchmark: an emphasis on integrated management, recognition of water's economic value, stockholder participation in decision making at the lowest appropriate level, access to water services by the poorest users, an ecosystem approach and private sector contribution. The most evolutionary of these principles, and the one cited more often, is perhaps the one pertaining to water as an economic good. To many, this evokes images of businesspersons and markets and of private sector involvement in the provision of services such as electricity, irrigation and water supply and sanitation, which present generations associate with the public sector and the common good.

However, the application of this principle is not new. Water services were provided by private parties in the United States and Europe before the 20th century, and in many countries it was not until the 1920s, 1940s and even the 1950s that these services were absorbed by local, state and national governments (Lee, 1990).

To cite an example from a developing country, the water supply in certain areas of Guatemala City was mainly privately provided until a Municipal Water Utility (EMPAGUA) was created in the 1960s. Moreover, a water market existed in Guatemala City until some 20 years ago. Whenever EMPAGUA wanted to increase the supply by tapping a new source, the Central Bank would provide finance and would then issue water titles. Anyone could buy one of these titles or 'Pajas de Agua' ('straws', named after the manner in which the aqueduct was taped in colonial times). They could be used to receive a volume of water equivalent to $2\,m^3$/day, or they could be negotiated or used as collateral, or they could simply be saved, waiting for prices to rise. When the titles from a given water supply source were sold out at the Central Bank, private buy-and-sell transactions occurred. This market collapsed, however, when the 'Pajas de Agua' could no longer be backed up any with the corresponding water volumes, due to increasing population growth rates that outpaced the increase in water supply. If people were willing to trade water, nobody wanted to buy paper.

Another interesting example is that of the Panama Canal, which has operated hydraulically based on water volumes regulated by two major reservoirs, which also provide water supply for municipal and industrial uses of the central region of Panama, including Panama City. Recently, the effects of El Niño and projections of future traffic increases and larger ship sizes prompted the Panama Canal Authority to think about additional sources of water both inside and outside the original Panama Canal watershed.

As the cost of the possible future inter-basin water transfers to supply the canal and the central Panama municipal and industrial water requirements are quite high, the Panama Canal Authority is studying all possible alternatives to meet the future requirements, including demand management. Among these, a new water pricing system for the use of water from the Panama Canal watershed is being studied. The main objectives of the new system would be the long-term conservation of the resource and to increase revenues for its administration and management.

Despite examples such as these, the public era of provision of services was the norm until the concept of water as an economic good was rediscovered in the late 1980s and emphasized by organizations such as the World Water Partnership and the multilateral and regional financial institutions in the late 1990s.

Present Trends in Latin America

The Inter-American Development Bank[2] (IDB) is the oldest and largest of the regional multilateral development banks. Established in 1959 to foster economic and social development in Latin America and the Caribbean, its membership currently totals 46 countries, 26 of whom are borrowing member countries.[3]

During four decades the Bank provided substantial financing to its borrowing member countries for all kinds of water-related infrastructure projects, namely water supply and sanitation, irrigation, drainage and flood control and hydroelectric generation, almost exclusively to the public sector. In the mid-1990s however, new trends in water resources management became evident in Latin America.

The emphasis in centralization for the provision of water services, which for many decades was promoted by many financial organizations seeking stronger institutions and economies of scale, was reversed and decentralization became a rediscovered paradigm. The application of economic instruments and the involvement of the private sector in the provision of water services were strongly emphasized, encouraged by the successful example of Chile and a perceived failure of the public sector to generally provide for these services efficiently.

This, and the recurrent economic crises that increased the difficulties for public financing of the provision, operation and maintenance of water services, called for a change in the role of the public sector from provision to regulation of such services. It was also a factor in the call for a change of paradigm from development of the supply to management of the demand and from a sectoral to an integral approach in the management of the resource. With all these changes in the water sector, it also became evident that the institutional framework that in the past had the responsibility for solving the water issues and the water problems of the population had to change and modernize.

In May 1998, the Bank approved its strategy for integrated water resources management (IDB, 1998). Perhaps the most emphasized of the guiding principles of the Bank's strategy relates to the institutional innovation and capacity building required to bring into practice these universal principles for integrated water resources management (IWRM), which were so clearly enunciated in Dublin.

There is a wide variety of institutional arrangements that go from market mechanisms for the allocation of water among competing uses, such as in Chile, to river basin organizations and from a negotiated bottom-up approach, such as in Brazil, to a regulated top-down approach; from public ownership to private ownership and from centralized authorities to decentralized entities, such as in Colombia. A summary of the situation and trends regarding the institutional framework for IWRM in Latin America and the Caribbean at the beginning of the millennium can be found in sources such as the paper by García (2002).

Generally speaking, countries do assign priorities to various uses. Water for human consumption gets first priority in most cases, but not in all. There are cases where water for hydroelectric generation has first priority and cases where use of water for livestock has a high priority, second only to the use of water for human consumption.

In the last decade, institutional changes in the water-use sub-sectors, such as water supply and sanitation, have taken place. The separation of the functions of providing water services, from that of regulating the provision of such services, has been the cornerstone of these changes. A new type of sub-sectoral water organization has thus appeared: the 'independent' regulatory entity.

However, this water-use institutional change paid scant attention, at least at the beginning, to the fact that what was being restructured and regulated was the provision of services, i.e. the use of the water, but not the management of the water as a resource. It was simply assumed that the supply would always be there and that the return flows would always find an adequate mean of disposal. Apparently, water use conflicts both in quantity as in quality and conflicting demands were not part of the equation, at least not at first.

This situation provoked the emergence of a new model of water resources organization, this time not to regulate the relationships between the providers of services and the users of these services, but to regulate the allocation of water resources among competing uses (García & Valdes, 2000).

The objective for this independent regulatory entity would be to reach an integrated management of the resource, by allocation of water use rights, including those for return flows. The criteria would be economic efficiency, financial sustainability and social equity, so that supplies could be ensured for all alternate uses and avoid conflicts, especially under scarcity conditions.

The creation of a regulated (or not) 'water use rights market' could also be considered. Water for human consumption and environmental needs could (or not) be excluded from the market, and these would have (or not) priority of use. All other uses would compete in this market. If necessary, this mechanism could be established by river basin.

Variations of this model have been favoured in actual operations by multilateral financing organizations such as the World Bank and the IDB (Van Hofwegen & Jaspers, 1998; ICWS/CRESEE, 2000; IDB, 1998; 2000a,b; World Bank, 2000).

Questions from an Outsider's Point of View

At present it would be hard to find a country where its water policies and/or strategies, if they exist, do not expressly endorse the Dublin principles. One of those principles is that water is an economic good, therefore, it must have a value. But what does this mean, in practical terms, for the water organizations and the consumers? From an outsider's point of view, not privy to the internal intricacies of charging for water, some degree of naiveté is, hopefully, permissible. Therefore, it is posed that the bottom line is that water must be valued and charges must be levied for its use. Additionally, from a practical point of view, posing the following question as the name of the game does not imply great difficulty: who charges who, for what, how, and how much?

One of the underlying premises of the institutional model described in the previous section is that the water resources regulatory entity is self-sustained: that it generates revenues from the allocation of water rights to the users only for its operation expenses. This is equivalent to providing a service and charging for it. Once a water right is allocated to a user, the market will assign the water to the most productive use, such as in the Chilean model. However, having been forced to look more closely into this, the naïve outsider cannot help but experience certain uneasiness, feeling that much is missing from the picture and that in the real world things are not that simple. There are too many actors and variables to consider and many more questions, such as the following, come to mind: Will the 'service charge' of the regulatory entity be equal for all users? If not, how and what to charge to different users and on what basis?

- If no economic instruments, such as a water rights market, are used for the allocation of water among competing uses and users, how much and how to charge to promote efficiency?
- If only a 'service charge' is made, does it mean that once the rights are obtained, water use is free? Should it be free or should the government charge again for its use, given that it belongs to the nation? If it were private property, the owner would surely charge.
- Should the product of these charges be considered as a source of national revenue?
- Should service providers or 'resellers' of water be charged as individual users? If that should not be the case, how and how much to charge each one?
- How much should service providers or 'resellers' charge their 'clients' and how?
- Should pricing strategies be universally designed by marginal cost pricing?
- Should pricing strategies be designed only for surface waters, in the absence of groundwater abstraction and pollution charges, or should a more comprehensive pricing strategy considering these be designed?
- Is there a 'right' price for water?
- Water has a value in-use, but its use also has a cost. Certain schools of thought promote charging the 'full cost of water', but what is it?

- The literature mentions the economic cost, which is easier to determine than the environmental externalities. Components of the economic cost are capital costs, operation and maintenance, opportunity costs and economic externalities (Bhatia *et al.*, 1999).
- Can pricing at 'full cost' or 'full pricing' be achieved in practice and under what conditions? Earlier proponents of this principle are now having second thoughts in the face of income limitations of large population groups and decreasing agricultural revenues in developing countries.
- Under these conditions, does its application reduce or increase equity?
- If the value of using water to the users is greater than the cost of using it, it makes sense to use it for that purpose, although there may be cases where the cost is greater than its value in-use.
- Does water have an 'intrinsic' value, even if it is not used by humans that should be added to its value in-use, and therefore included when determining its 'full price'? Although there is a school of thought favouring this, there does not seem to be a consensus about how to assess this intrinsic value.
- Will different pricing schemes produce different consequences for integrated water resources management?
- Should pricing strategies be tailored to the circumstances at hand, to influence and change water managers' and users' behaviour?
- Are certain policy results achievable through water pricing schemes?
- Is water pricing a policy tool or only a revenue tool?

Some Difficulties Involved in Water Pricing

The fact that water is also a basic necessity for living and that it also has a social and environmental value, makes it extremely difficult to select an appropriate price (IWMI, 1998). Sometimes water is a private good and sometimes it is a public good. In some cases the marginal utility of water is very high, such as in a drought for irrigation purposes, and sometimes it is very low, making it difficult for markets to function. In addition, given the complexity of the flow of water through a river basin, transaction costs may be high, and insecure and undefined property rights and high incidence of externalities are major causes of market failures in developing countries (Perry *et al.*, 1998).

Since water is a basic human need, cultural values and value judgements have important implications. Many people believe that there is an obligation for society to assure reasonable levels of water, the same as food, shelter and medical care up to a certain level, to ensure that basic human needs are met, but not beyond, suggesting that water has a different value for different levels of consumption (IWMI, 1998). Because of this, among other reasons, some believe that water policy decisions must be formulated in terms of multi-objective decision making, recognizing that the relevance and importance of the different values will vary under different conditions of time and place (Perry *et al.*, 1998).

Some Useful Guidelines

Hanemann (1999) presents an interesting discussion about the role of pricing in promoting integrated water resources management from a US perspective. Bhatia *et al.* (1999)

discuss the use of prices to promote equity, efficiency and sustainability, drawing from experiences in India and Jakarta and OECD countries. What follows is summarized from these discussions.

Hanemann (1999) takes a pragmatic approach and poses that a pricing strategy should not be designed in isolation by a mechanical application of marginal cost pricing. It needs to be tailored to the circumstances at hand, since the underlying goal of water pricing is both to influence and change water managers' and water users' behaviour. In what direction it is desired to influence this behaviour, and how this can best be accomplished, requires local judgement and will vary with circumstances.

For example, in an illustrative manner Hanemann discusses four major and two subsidiary possible objectives for designing a rate structure. The major objectives are:

- Raising revenue (for financial sustainability, stability and predictability).
- Allocation of costs among different uses and users (for social equity, political acceptability).
- Changing behaviour (by providing effective incentives to users).
- Promoting economic efficiency (both in the use of water as well as to new investments).

The subsidiary objectives are:

- Ease of administration (transparency, simplicity).
- Avoiding negative environmental externalities and promoting environmental sustainability).

Rate design depends on the objectives and these vary with circumstances. Sometimes the focus is backward looking, to raise revenue in order to recover past costs. Emphasis is on covering existing financial obligations and to provide water as inexpensively as possible. Future financial obligations will be dealt with later. This is the objective of raising revenue (objective 1) and Hanemann (1999) poses that it is reasonable as long as it does not conflict too much with the objective of allocation of costs among different uses and users (objective 2) and the same purpose cannot be attained through marginal cost pricing (objective 4). Additionally, an adequate supply of water must exist and there should not be a foreseeable need for future investments in water supply. If either one of these fails to hold, a forward-looking approach is proposed to change water behaviour (objective 3) or promoting efficiency (objective 4).

It is easy to see that these objectives may have conflicting implications for rate design (Hanemann, 1999). For example:

- Reducing what users pay attains the objective of raising their welfare but diminishes the incentives for efficient use of water and lowers revenue.
- Fixed charges promote stability and predictability, but diminish the incentives for efficient use of water.
- Charges based on consumption promote conservation and efficiency in water use, but cause uncertainties in the revenue stream.

The conclusion is that no single approach can be globally applied, but depends on the objectives and the circumstances. Along the same lines, Bhatia *et al.* (1999) state that there is no universal 'best' rate design. What is 'best' for a given community and situation is that which reaches an acceptable compromise among the objectives that are important for that community (Rogers *et al.*, 2001). In a more extensive and closer to reality manner, they discuss not six but 17 objectives. It can be imagined that the situation and the implications become even more complex in the real world.

In the case of water use tariffs, OECD countries use some combination of the following elements (Bhatia *et al.*, 1999):

● Connection charges
● Fixed charges
● Volumetric charges
● Block charges
● Minimum charges

For example, some countries use a two-part tariff system, which has fixed and variable elements. Other countries include an increasing block tariff system for the variable part, providing different prices for two or more pre-specified blocks of water, raising the price for each successive block. In this case, decisions must be made on the number of blocks (a managerial decision), volume of water associated with each block and price to be charged for each block (these last two are social and political).

Moreover, there is more than one method for setting the tariff prices in relation to costs, such as average incremental costs, marginal cost, price cap, benchmark or 'Empresa Modelo', and average referential price (Bhatia *et al.*, 1999). Elasticities must also be considered and there are also groundwater extraction charges, wastewater and sewerage charges and subsidies to account for, making tariff design also a complex endeavour.

As an example, the Venezuelan experience in setting the tariff structure for Caracas is explained by Rubinstein (1999). In a highly discretional environment lacking regulation, Hidrocapital, the Caracas' water utility was motivated by two factors: (1) government fund transfers ceased and (2) an intended privatization process was aborted. The objective was to increase revenue to cover costs and several methodologies were explored, finally the average referential price method was selected.

As a background, the rate setting procedure prior to 1993 was long and complex and highly political. It involved at least seven consultation and approval instances with various ministries, with the possibility of re-negotiation loops. After 1993, a simplified procedure was allowed, enabling the water utility to set the rates using inflation or its best estimate as a proxy for increase factor and deciding the frequency of rate increases. Two parameters set the price: an average referential price and an average referential price for social users. Cross-subsidies between users were allowed and it was automatically indexed over time.

At the time, the water sector in Venezuela was enjoying a regulatory vacuum, not because of laws and regulations, and not because of lack of actors, but due to lack of direction and leadership, fragmented and dispersed responsibilities, and lack of penalties for non-compliance (Rubinstein, 1999). This vacuum gave the water utility the ability to increase the tariff and to explore different ways of private participation and shielded it from political interference. However, as a disadvantage, the rules of the game were unclear, producing a highly discretional, unstable and risky situation, and responsibilities and accountability were not defined. Even then, the conclusion was that lack of regulation was better than inappropriate regulation and that the tariff increase improved the financial situation of the utility, without a significant consumption reduction (Rubinstein, 1999).

Conclusions

From an outsider's point of view, it is evident that there are many questions and no easy answers. It is very clear, however, that determining the value of the use of water and water pricing is a complex process.

It is also clear that when talking about water pricing, wholesaling and retailing, and charging for treated water as well as charges for 'raw' water should be clearly differentiated and should not be lumped together or confused. Therefore water pricing goes beyond, and should not be confused with, the determination of water utility tariffs, a process equally complex.

Water pricing is an important policy component of water resources planning and management that, because of its complexity, is often not considered by water resources managers. Nevertheless, it is an important component of water resources management and planning and more attention should be given to it in all of its possible implications and objectives, including influencing and modifying behaviour, and not only to the objective of increasing efficiency.

It is also evident that there is no universal model. The solutions have to be tailored for each case and be acceptable by the community.

In Latin America water pricing has been traditionally considered mainly on a retail scale with a user welfare objective. In some instances, the objective has been shifted towards increasing the revenue of the water utilities. Water pricing on a wholesale scale for 'raw' water is less frequent although it is duly being considered in some of the large countries such as Mexico and Brazil.

Notes

[1.] International Conference on Water and the Environment: Development Issues for the 21st Century, 26–31 January, Dublin, Ireland.
[2.] Throughout this paper the Inter-American Development Bank will be referred to as the IDB or the Bank.
[3.] Argentina, Bahamas, Barbados, Belize, Bolivia, Brazil, Chile, Colombia, Costa Rica, Ecuador, El Salvador, Guatemala, Guyana, Haiti, Honduras, Jamaica, Mexico, Nicaragua, Panama, Paraguay, Peru, Dominican Republic, Suriname, Trinidad and Tobago, Uruguay and Venezuela.

References

Bhatia, R., Rogers, P. & de Silva, R. (1999) *Water is an Economic Good: How to Use Prices to Promote Equity, Efficiency, and Sustainability* (Cambridge, MA: Harvard University).

García, L. E. (2002) Institutional framework for Integrated Water Resources Management in Latin America: some experiences from the Inter-American Development Bank. International Water Resources Association, *Proceedings of the IWRA Policy and Institutions Workshop, Salvador Bahía, Brazil, September 3–7, 2000* (Oxford: Oxford University Press).

García, L. E. & Valdes, J. B. (2000) Water resources sustainability for the next millennium—The Latin American case, in: U. Maione, B. Majone Letho & R. Monti (Eds) *New Trends in Water and Environmental Engineering for Safety and Life* (Rotterdam: Balkema).

Hanemann, M. (1999) The role of pricing in water resources management, *Special GWP Seminar in Water Pricing. Stockholm, Sweden* (Berkeley, CA: Department of Agricultural and Resource Economics, University of California at Berkeley).

ICWS/CRESEE (2000) *Marco Institucional y Legal para el Manejo Integrado de los Recursos Hídricos en Costa Rica*, Informe Final. Cooperación técnica ATN/NE-6333-CR (San José: ICWS/CRESEE).

IDB (1998) *Strategy for Integrated Water Resources Management*, Inter-American Development Bank, Sustainable Development Department, Environment Division. IDB Strategy Paper No. ENV-125 (Washington DC: IDB).

IDB (2000a) *Estudio del Marco Legal e Institucional para el Manejo de los Recursos Hidricos de Paraguay*, Inter-American Development Bank, Informe de consultoria DBEnvironment, Cooperacion Tecnica ATN/FC-6006-PR (Asuncion, Paraguay: IDB).

IDB (2000b) *Hacia un Manejo Integrado de los Recursos Hidricos en Honduras*, Inter-American Development Bank, Informe de Consultoria IHE/Delft and Resource Analysis, Cooperacion Tecnica ATN/NE-6331-HO (Tegucigalpa: IDB).

IWMI (1998) Water as an economic good, *News*, 2(1) (Colombo, Sri Lanka)

Lee, T. (1990) *Water Resources Management in Latin America and the Caribbean*, Studies in Water Policy and Management, No. 16 (Boulder, CO: Westview Press).

Perry, C. J., Seckler, D. & Rock, M. (1998) *Water as an Economic Good: A Solution or a Problem?*, IWMI Research Report No. 14 (Colombo, Sri Lanka: IWMI).

Rogers, P., Bhatia, R. & Huber, A. (2001) *El agua como bien económico y social: como poner los principios en práctica*, TAC Background Papers No. 2 (Santiago: Global Water Partnership (GWP)).

Rubinstein, J. (1999) The Venezuelan experience: successful tariff structure in Caracas 1993–1999, Special GWP Seminar in Water Pricing, Stockholm, Sweden.

Van Hofwegen, P. J. M. & Jaspers, F. G. W. (1998) *Analytical Framework for Integrated Water Resources Management—Guidelines for Assessment of Institutional Frameworks*, IHE Monograph 2 (Rotterdam: IHE).

World Bank (2000) *Argentina, Water Resources Management: Policy Elements for Sustainable Development in the XXI Century. Direccion Subregional Argentina, Chile y Uruguay* (Washington DC: The World Bank).

Water Pricing Reforms: Issues and Challenges of Implementation

LUIZ GABRIEL T. DE AZEVEDO & ALEXANDRE M. BALTAR
World Bank, Country Office, Corporate Financial Center, Lote A, Brasilia-DF, Brazil

Introduction[1]

The scarcer a resource becomes, the more efficient should be its use. Water resources do not follow this rule. Limited renewable availability and increasing demands have made scarcity of water resources one of the world's most challenging issues with serious local, national, regional, as well as global impacts.

The realization that water is finite led to the development of new concepts and approaches in an attempt to effectively address the multiple challenges facing water resources management. The scope of a new framework has been widely debated in a number of important international events during the past two decades resulting in a broad global consensus, forged during the Rio Earth Summit process, which recommends that modern water resources management be based on three fundamental principles (known as 'the Dublin Principles'). First there is the ecological principle, which argues that water management has to be comprehensive, integrated and inter-sectoral, that the river basin must become the unit of analysis, that land and water need to be jointly managed, and that much greater attention needs to be paid to the environment. Second is the institutional principle, which argues that water resources management is best done when all stakeholders participate, including governments, the private sector and civil society; that women need to be active participants of the decision-making process; and that resource management should adopt the guiding principle of subsidiarity, with actions taken at the lowest appropriate level. Third is the instrumental principle, which argues that water is a scarce resource, and that greater use needs to be made of incentives and economic principles in improving allocation and enhancing quality.

It is within this context that water pricing ought to be considered. The use of pricing mechanisms is not an objective in itself but an effective means of improving water resources management and the provisions of water services. Therefore, the implementation of water pricing reforms should always be evaluated within a broader perspective that takes into account elements related to the three principles, ecological, institutional and instrumental, as well as political, social, natural, legal and cultural characteristics of each country or region.

It is also important to realize that even when conditions are favourable, water pricing reforms are quite difficult to implement. Although significant progress has been achieved in many countries, these can be regarded as pilots or examples in comparison with the huge challenges that still remain. This is a process that requires persistence and patience. As stated in the World Bank Water Resources Sector Strategy, "all countries, including industrialized ones, have a long way to go before they manage their water resources in accordance with principles of best practice. The challenge of reform, accordingly, is to determine what is feasible, in any particular natural, cultural, economic and political environment, and to develop alliances around a sequenced, prioritized, realistic program for improvement" (World Bank, 2002).

This paper deals primarily with this process of improving the implementation of water pricing reforms.[2] It is primarily based on the World Bank experience in supporting such reforms in developing countries. In the sections that follow the objectives, importance, critical challenges, options and recommendations for the implementation of water pricing reforms will be discussed.

Why Water Pricing Is Important

The World Commission on Water (WCW) has estimated that to meet all water supply and sanitation, irrigation, industrial and environmental management demands, investments in water infrastructure need to increase from the current level of $75 billion to $180 billion a year (World Water Vision, 2000, p. 51). This enormous investment gap will demand innovative thinking and that co-operative approaches be met, while well-targeted subsidized public investments will still be needed. The development and long-term sustainability of the necessary infrastructure will certainly require the systematic adoption of integrated water resources management and introduction of appropriate water pricing mechanisms.

Without adequate pricing mechanisms, consumers have no incentive to use water more efficiently as they receive no signal indicating its relative value (Asad *et al.*, 1999). Similarly, if the water service providers are unable to recover the costs to adequately fund their operation, systems will deteriorate and the quality of service will suffer. The deterioration of water systems can be seen worldwide, particularly in developing countries. Finally, if the cost of maintaining clean water is not incorporated into prices charged to relevant users, then there will be little incentive to reduce water pollution. As a result, freshwater supplies will become increasingly unsafe.

Free or under-priced resources are frequently misallocated, mismanaged and wasted. The WCW Report stated that "where services are free, the result is inevitably politicization of the concerned agencies, inefficiency, lack of accountability, capture of the subsidies by influential groups, and a vicious cycle of poor quality services, water rationing, and insufficient resources for operation, maintenance, and investment. In almost all cases, the poor end up without access to water. They also pay exorbitant prices for inferior and unreliable services provided by unregulated vendors and bear the brunt of environmental degradation" (World Water Vision, 2000, p. 34).

To summarize, the introduction of adequate water pricing reforms gives rise to a series of fundamental and healthy changes, being essential in the process of breaking the vicious cycle outlined above, by: (1) giving users a sign of the economic value of the resource, thus helping to promote its more efficient use; (2) providing financial resources to guarantee adequate administration, operation and maintenance (A, O&M) of water infrastructure; and (3) funding (at least partially) water resources management and development.

Why Is It so Difficult to Implement Water Pricing Reforms?

The World Bank has recently conducted a detailed assessment of the implementation experience of its 1993 Water Resources Management Policy, taking into account more than 400 Bank-financed projects in over 100 countries (World Bank, 2001). This evaluation came to the conclusion that "there was a general failure to promote compliance with cost-recovery provisions and the implementation of water tariffs and charges that reconciled economic efficiency, social equity, financial criteria, and autonomous and independent regulatory systems" (p.1). Indeed, 78% of countries participating in sector operations have failed to comply with covenants agreed with the Bank regarding this matter. One of the reasons identified for this mediocre performance was that the Bank may have required reforms to be implemented too quickly or before the borrowers were fully committed.

This demonstrates that even with important external technical and financial support, water pricing reforms are extremely complex and take time to mature. Industrialized countries provide another example. A detailed recent review by the Organization for Economic Cooperation and Development (OECD, 1998) shows that even in the most advanced countries, where sound governance, participation, institutions, and skills do exist, much progress remains to be made on issues such as: (1) "more consistent application of the polluter pays principle"; (2) "prices very rarely reflect full economic and environmental cost"; (3) "most work in improving water use efficiency remains to be done"; and (4) "agricultural water use still heavily subsidized". Still, the OECD report stated that, while many notable successes have been achieved, "the progress to date is the result of many years of effort" (OECD, 1998, pp. 29–30).

Indeed, water is not at all an easy-to-manage resource. There are some special features, such as the need for capital-intensive infrastructure, public good nature of certain water activities, and occurrence of third-party impacts (see Box 1), that add significant complexities to the financial equation of water resources management.

There are also other conceptual and practical issues that make the implementation of water pricing reforms even more difficult. These include: (1) the dual objectives of water pricing—economic efficiency and cost recovery; (2) different perceptions by economists and users on what should be considered appropriate pricing; (3) the inherent complexity of the irrigation sector; (4) the lack of accountability in water service provision; and (5) difficulties in taking opportunity cost into account.

Box 1 Special characteristics of water

- Large, lumpy capital requirements and economies of scale in water infrastructure tend to create natural monopolies, warranting regulation to prevent overpricing. Moreover, many water investments produce joint products, such as recreation, electric power, flood control and irrigation, which make pricing and allocation decisions difficult.
- The large size and extremely long time spans of some investments, given under-developed capital markets and the potential for political interference in many water infrastructure investments, reduce the incentives for private investments in the sector; in such situations, public investments may be warranted.
- The uses of water within a river basin or aquifer are interdependent. Withdrawals in one part of the basin reduce the availability of water for other users; groundwater pumping by one user may lower the water table and increase pumping costs for all users; and pollution by one user affects others in the basin, especially those located downstream. These interdependencies suggest that having all users agree to the rules of the game, or lacking that, imposing government regulations, taxes, or both, could improve the social value of water resources.
- Certain aspects of water activities, such as the control of floods and waterborne diseases, are (local) public goods, which cannot easily be charged for on the basis of individual use. In such cases, public initiative may be required to ensure that levels of investment are appropriate.
- Water resources are often developed because of their strategic importance for national security and for regional development. Governments thus typically maintain ownership of water thoroughfares, providing services such as the coast guard and traffic regulation.
- Some regions are subject to periodic droughts. Because water is essential to sustaining life, governments may take control of water.

Source: World Bank (1993), p. 28.

Economic Efficiency versus Cost Recovery

Some economists have argued that the ideal theoretical solution for water pricing is to set economic efficiency as the main objective and set prices according to full economic cost recovery criteria. Attaining this ideal, however, is generally not practical. Administrative and/or information complexities and political realities need to be taken into consideration, thereby yielding a more attainable objective. Setting prices according to the objective of economic efficiency, although a valid and foreseeable goal, requires detailed information on opportunity costs and value of water in alternative uses, which is at best difficult and expensive to obtain and very often provides misleading signs for policy making and implementation (Azevedo & Asad, 2000).

The Brazilian case is a good example of how pricing objectives differ in theory and practice. Although many studies have been produced in different states and river basins looking at estimates of optimum prices that capture economic values of alternatives uses, the two most relevant initiatives in Brazil relied, instead, on values decided through broad political negotiations with main water users. This is the case in the state of Ceará in the dry and poorer Northeast region, the only state where bulk water is actually being charged in Brazil; and the case of the Paraíba do Sul River Basin, an industrialized and developed basin shared by the states of Rio de Janeiro, São Paulo and Minas Gerais, where the introduction of bulk water pricing has been recently approved by the River Basin Committee. In such initiatives, agreed prices were lower than optimum values from the standpoint of economic efficiency, but can still stimulate a more efficient behaviour by water users and provide financial resources to promote adequate and sustainable water resources management and development. In other words, even with prices set for the primary objective of recovering costs, there has been an efficiency enhancing effect due to the simple fact that there was no payment at all before the reforms took place.

It is also clear that there are different definitions and levels of cost recovery. As normally happens to any capital-intensive infrastructure, some amount of subsidy may be required to form the investment package. Accordingly, charges may be set at a level to achieve full cost recovery (i.e. including a lease payment that is designed to amortize the cost of the investment over its expected life) or merely operational cost recovery. This will depend upon the objectives of the project and the nature of any subsidies that have been provided.

Since a failure to operate and maintain new infrastructure does not have immediate visible effects, potential or actual water users may argue that zero or minimal charges will encourage a rapid take-up of the newly available resources. Public bodies responsible for the construction or regulation of the infrastructure will be strongly inclined to waive charges or set them at a level well below the long-term cost of operations and maintenance. This makes economic sense in the short term, provided that charges are increased as demand builds up. Unfortunately, the expectations created by low initial charges are easily converted into a quasi-property right with users believing that they should never be required to pay realistic charges for their water. This is an example of the time-inconsistency of a policy strategy that can often contribute to derailing the pace of pricing reforms.

What Is Considered Appropriate Pricing?

The second mentioned issue refers to the different perceptions that economists and water users generally have with respect to what should be considered appropriate pricing. Economists argue that users should be charged for the marginal costs of producing the next unit of the resource. Users, on the other hand, understand a price as a payment for a service rendered and, as this service is often provided in a regime of monopoly, a legitimate price for users is that which covers the costs of an efficient provider to provide the service. Thus, users consider average, not marginal, cost to be appropriate.

Additionally, there is no consensus with respect to what cost components should be included in the price to be charged to users. Users consider legitimate payment for the costs of operating and maintaining the existing infrastructure and, through transparent negotiation and effective communication, also the costs of replacement. However, they vigorously resist the notion that they should pay for sunken costs, which, in their eyes, have already been paid for by taxes or other assessments.

Similarly, the costs of water resources management that are associated with public good-natured activities, such as flood control, drought alleviation and environmental quality, cannot be fully charged to individual users. These costs are often diluted within the overall costs of water agencies and, since they are not easily separable, their inclusion in the price to be paid by users becomes another issue to be negotiated.

Inherent Complexity of the Irrigation Sector

The third issue listed above relates to the complexity of the irrigation sector. The agriculture sector accounts for about 70% of all fresh water used in the world. Figures vary greatly among countries, but almost everywhere irrigation is the largest water user.

Experience shows that this sector is also one of the most active opposers to water pricing reforms. Generally, the stronger the political influence of agricultural sector the stronger will be the opposition to water pricing reforms.

In such a context, a basic difficulty with most pricing mechanisms that are designed to guarantee adequate resources for A, O&M and better utilization of the existing infrastructure is that these are open to the charge of discriminating against small farmers. From a political perspective this may be a powerful argument even though it is based on the most superficial analysis of the impact of the changes. Many studies as well as practical experience provides ample evidence that existing arrangements around the world tend to benefit large water users over small-poor farmers. When users are not required to pay for the use of water, large and influential users are usually able to ensure that they receive better service—greater reliability or higher volumes—than other users. Thus, large users are likely to have relatively less to gain from the adoption of charging and other arrangements designed to sustain better management. Small users, even where they should gain over a number of years, may be understandably reluctant to support a change which brings the certainty of higher cash payments combined with less certain promises of better services and higher incomes. The complexities and challenges of overcoming these perverse incentives to maintaining the status quo should not be underestimated.

Second, it should be considered that there are fundamental distinctions between the markets in which irrigation and other water users (e.g. urban water supply) operate. While in the case of urban water supply the end product can be considered as a local non-tradable good, in the case of irrigation, where the end products are agricultural goods which trade on a global market, the situation is radically different. For example, as the agricultural subsidies from OECD countries are huge (about $ 350 billion/year) (World Bank, 2002) this has major impacts on the prices of agricultural products in developing countries. This is (reasonably) used by farmers as an argument against the introduction of water charges.

Lack of Accountability in Water Service Provision

What happens if the service provider is not efficient? Experience shows that most attempts to increase cost recovery tend to fail if the accountability issue is not resolved, since users will obviously and legitimately have no incentives to pay for the cost of water provision given the inefficiencies of the service providers.

In such situations effective progress in the implementation of pricing reforms requires more than the simple exhortation of the benefits and importance of cost recovery to improving water resources management and/or environmental quality. Instead, it is necessary to realign institutional arrangements and incentive structures in such a way as to ensure that water suppliers are accountable to users and that water charges become a real tool guiding service provision.

Besides its potential to hinder or even to cause the failure of cost recovery efforts, the lack of accountability in water services provision has many other negative consequences for the management of water resources. In the water supply and sanitation sector, for example, experience shows that unaccountable, financially unsustainable utilities often operate in a climate of impunity, paying little attention to public responsibilities, such as wastewater goals and water quality standards set by river basin authorities and other public water resources management agencies. Conversely, accountable utilities are acutely aware that their business depends on the availability of a reliable quantity of good quality water.

With that in mind, they understandably become active advocates of effective water resources management.

Difficulties in Taking Opportunity Cost into Account

Should prices reflect the opportunity costs of water in each one of its alternative uses? The right answer seems to be a questionable 'yes' in theory and a positive 'no' in practice (see Box 2). Theoretically, "in an economically efficient resource allocation, the marginal benefit of the employment of the resource is equal across uses, and thus social welfare is

Box 2 Financial versus opportunity costs (practical implications)

"User payments for the financial costs of services rendered is a fundamental requirement for any financially sustainable water supply system. This is very important. But the claims for 'pricing' typically go beyond that of maintaining and operating infrastructure, and suggest that if 'the prices are right, allocation will be optimal'.

Proceeding from the point of view of users (as one must when considering political economy of reform rather than theoretical elegance), it is vital to distinguish between two radically different types of cost. First there are the costs that any user can understand, namely the financial costs associated with pumps, treatment plants and pipes. Second is the far more subtle concept of the opportunity cost of the resource itself. There have been many proposals (almost all of which have come to naught) for doing sophisticated calculations of this opportunity cost, and charging users for this 'to ensure appropriate resource allocation'. This has not worked for two fundamental reasons. First, because it is impossible to explain to the general public (let alone to angry farmers) why they should pay for something which doesn't cost anything to produce. And, second, because those who have implicit or explicit rights to use of the resource consider (appropriately) such proposals to be the confiscation of property. (It is equivalent to asking a homeowner to pay rent at the market rate to the government for occupying the property which she owns.)

An added, and very important, factor is that the ratio between financial and opportunity costs is often radically different for different sectors. Although everything in water (like politics) is local, there are two broad patterns. It costs a lot to operate the dams, treatment plants, pumps and pipes which provide households with the modest amounts of water they use. Alongside these large financial costs, the opportunity cost of the resource itself (as measured by the value of the raw water in its next best use, often irrigation) is typically quite low. For municipal and industrial water, therefore, financial costs generally dominate opportunity costs.

For irrigation the situation is almost exactly the opposite. It costs relatively little (per unit of water) to build, operate and maintain the usual gravity systems which provide very large quantities of water. But the opportunity cost of the water (for cities and, increasingly, for high-value agricultural uses) is, in situations of scarcity, often much higher (a factor of 10 in typical cases of scarcity) than the financial cost of supplying the water.

These numbers (remembering, of course, that every place is different) have profound implications. They mean that, from the point of view of ensuring that users take into account the cost of the resources they are using, the emphasis must be on financial costs for municipal supplies, and on opportunity costs for irrigation. (Cost recovery for irrigation remains, as discussed above, very important for infrastructure sustainability, but not for allocative efficiency, which is the focus in this section.)

The great challenge for irrigation, then, is how to have farmers take account of the opportunity cost of the resource when it is both wrong in theory and impossible in practice to charge them for this."

Source: World Bank, 2002, p. 25.

maximized. This equilibrium can be achieved through the operation of price signals in a competitive marketplace for the resource" (Gibbons, 1986, p. 2). Only in this ideal case, opportunity costs would be clearly obtainable. However, in the absence of an operating market, how reliable would be the estimates of these opportunity costs?

In any event, and as shown in Box 2, in certain circumstances (such as irrigation) it would be necessary and fundamental to take account of opportunity costs in order to give users the right signal of the real cost of the water they are using. However, in such cases pricing mechanisms (charges collected by a water agency) are not the appropriate approach to follow. The best way to internalize opportunity costs into users' decisions would be the establishment of transferable water rights and creation of adequate conditions for water markets to evolve.[3] However, this is a controversial idea in many parts of the world, and in most countries legal frameworks and institutional arrangements are not sufficiently developed to allow for the broad use of market mechanisms. However, it should be remembered that as the challenges to water resources management increase it may be necessary to amplify the menu of alternatives available to water managers to include a combination of economic instruments such as pricing and markets, as well as other economic mechanisms capable of promoting more efficient management of water.

The Political Economy of Improvement

Experience over the past two decades indicates that the implementation of water pricing reforms is a complex process that often challenges long standing institutional, legal and cultural values. The consensus on the need to establish pricing reforms was forged a long time ago. Nonetheless, actual implementation of this consensus on the ground has been, at best, mediocre. It is now known that the international water community as a whole may have underestimated the challenges and the complexity of implementing such reforms especially under the vast variety of physical, climatic, historic, legal, cultural, institutional, etc., conditions around the world.

On the other hand, there have been many lessons learned during the past few years and these should be the basis for adjusting and revising strategies and developing additional instruments and mechanisms. Political economy is a key aspect of successful reforms and the experience thus far offers an indication of some of the aspects that need to be taken into account during the reform process:

- *Undertake reform only when there is a powerful, articulated and clearly recognized need for reform.* The perception of the need to reform cannot be exogenous and promoted primarily by external forces but rather endogenous and strongly recognized and felt by a large number of stakeholders.
- *Involve all users and other stakeholders in the process.* This is another principle that has proven to be much more difficult in practice than reaching theoretical agreement. Nonetheless, the effective involvement of users and stakeholders in the decision-making process is an essential requirement of a successful reform path.
- *Be aware that reform is dialectic, not mechanical.* The process of reform is not often associated with a linear path. It involves variables that are subjective in nature and for which mechanistic approaches are not very well suited.
- *Pay attention to general principles but adapt them to different institutional circumstances.* There is no 'one-size-fits-all' model or a magic formula that can be

applied everywhere and to every situation. Use should be made of principles which are broad in nature and that have been shown to be the basis for successful reform, but it must be taken into account that the implementation of such principles may require innovative thinking so that they can be adapted to different circumstances.

- *Start with the easier problems to build momentum.* The implementation of reforms should be sustained by positive results. For that to happen it is important to begin with simpler problems or situations for which solutions may be achieved within a reasonable timeframe. Once achieved, such successes would provide lessons and incentives to move forward into more complex challenges.
- *Acknowledge that there are no perfect solutions.* The use and/or allocation of water is a highly complex and controversial issue in most places. Very rarely, consensus can be achieved as the appropriation of the resource by one sector or by upstream users and often represents a less than optimal allocation by other sectors or downstream users. Therefore, it is essentially important to recognize that the implementation of reforms requires negotiation and acceptance of trade-offs that must be clearly understood by all interested parties.

Dinar (2000) provides an interesting discussion about the political economy of water pricing reforms and a detailed set of recommendations which would contribute to successful implementation of water pricing reforms, including:

(1) water pricing reforms should be launched after extensive public awareness campaigns;
(2) reformers should communicate a clear economic rationale, develop a broad reform agenda, adjust to institutional and political reality, and take account of traditional customs and social structures;
(3) successful reform programmes must include compensation mechanisms negotiated with stakeholders;
(4) reformers should precisely identify their objectives;
(5) reforms should be well prepared, because once they are implemented, they are hard to modify;
(6) the implementing agency must be sensitive to political events when putting the reforms in place;
(7) the agency should package and sequence the reform components to minimize opposition;
(8) it should seek external support and mobilize supportive stakeholders as much as possible;
(9) gains from reforms have to be shared;
(10) pricing reforms should acknowledge asymmetric upstream-downstream externalities, as well as the differences between water sources (groundwater and surface water);
(11) reformers should acknowledge the need for a set of institutions and not impose a generic process for reform implementation.

Conclusion

Experience to date indicates that reaching agreement on the strong relationship between water pricing and improved water resources management was much easier than the actual implementation of proposed pricing reforms that have proven to be more complex and difficult than originally anticipated. Despite a large number of studies and pilot initiatives only a handful of examples have been implemented where water and/or pollution charges are actually being collected. Nonetheless, the results of a vast number of initiatives, while stopping short of expected results, have provided substantial contributions to raising the level of the debate, to increasing our knowledge about the economics and many other aspects of water pricing. Many lessons have been learned and the water community is now better prepared to move forward.

Regarding the authors' experience with conceptual and practical issues that make the implementation of water pricing reforms difficult, five have been presented: (1) the dual objectives of water pricing—economic efficiency and cost recovery; (2) different perceptions by economists and users on what should be considered appropriate pricing; (3) the inherent complexity of the irrigation sector; (4) the lack of accountability in water service provision; and (5) difficulties in taking opportunity cost into account. The challenges, however, may not be limited to these five areas alone and other relevant aspects may have to be considered under specific settings.

One of the important questions considered in this paper relates to the objectives of water pricing reform. In this regard, one should not be afraid of adopting pragmatism as a course of action. In most settings, including developed countries, the most obvious objective influencing water pricing reforms is to recover costs.

While many have argued that the ideal theoretical solution for water pricing is to set economic efficiency as the main objective and set prices according to full economic cost recovery criteria, attaining this ideal, however, is generally not practical for several reasons. One of the main challenges is the estimation of opportunity costs for different water uses. Such costs are generally known to be excessively complicated and/or expensive to estimate at best, and completely misleading at worst. As such, increasingly, international experts are coming to the conclusion that allowing the market to determine opportunity cost prices is more sensible. This requires creating the conditions for water markets to evolve, which constitutes a gigantic challenge in itself.

A second practical challenge is that most bulk water agencies/companies are doing well if they can recover O&M costs and a portion of investment costs for bulk water supply. More typically, these services are partially or fully subsidized by public institutions. This is mainly a historical/cultural problem, which relates to the fact that in most countries water users are accustomed to paying little or nothing for bulk water. This, then, leads to a political difficulty, whereby policy makers and politicians are generally reluctant to adopt any bulk water pricing reform at all for fear of alienating powerful water user interest groups and/or individual users/voters.

At the political level, introducing water pricing reforms at all is quite challenging, particularly since most users are accustomed to paying little or nothing for water. For the above reasons, and others, it is preferable to set aside the ideal solution (at least initially) in favour of one which sets cost recovery as the main water pricing objective. Experience demonstrates that the role of pricing, with or without consideration of environmental costs, in raising the efficiency of water resources allocation and/or use has been generally secondary. On the other hand, application of cost recovery has provided for increased discipline in considering all costs and in reaching decisions regarding new investments. Finally, as discussed in this paper, the accountability and improved services provision to water users are essential elements of successful reforms. Users will not be lured into paying higher or new charge if this is not associated with perceivable benefits to them.

Regardless of the approach taken to implement water pricing reforms, it is critical to involve all stakeholders in the process. This includes both upstream involvement in the design of pricing schemes, as well as downstream in the implementation of the schemes, and collection and allocation of associated revenues. Once again, this has been a consensus that in practice has proven to be complex and challenging to implement.

Finally, the message of this paper is neither of challenging the conceptual framework regarding the importance of water pricing reform nor one of negativism regarding the potential for actually implementing the agreed principles. Rather, the authors sought to review the reasons behind a less than optimal implementation performance over the past several years with the objective of extracting lessons and suggestions that would lead to better results in the future.

Notes

[1.] The findings, interpretations, and conclusions expressed in this paper are entirely those of the authors and should not be attributed in any manner to the World Bank, or its affiliated organizations, or to members of its Board of Executive Directors or the countries they represent.

[2.] For discussions on specific country experiences, see Dinar & Subramanian, 1997.

[3.] Water markets are not discussed in this paper. For an overview of World Bank experience on this matter see Simpson & Ringskog (1997); Dinar *et al.* (1997); Briscoe *et al.* (1998); Marino & Kemper (1999).

References

Asad, M., Azevedo, L. G. T., Kemper, K. E. & Simpson, L. D. (1999) Management of water resources: bulk water pricing in Brazil, Technical Paper No. 432 (Washington DC: World Bank).

Azevedo, L. G. T. & Asad, M. (2000) The political process behind the implementation of bulk water pricing in Brazil, in: A. Dinar (Ed.) *The Political Economy of Water Pricing Reforms* (New York: Oxford University Press).

Briscoe, J., Salas, P. A. & Pena, H. (1998) *Managing Water as an Economic Resource: Reflections on the Chilean Experience* (Washington DC: World Bank).

Dinar, A. (2000) Political economy of water pricing reforms, in: A. Dinar (Ed.) *The Political Economy of Water Pricing Reforms* (New York: Oxford University Press).

Dinar, A. & Subramanian, A. (1997) Water pricing experience: an international perspective, Technical Paper No. 386 (Washington DC: World Bank).

Dinar, A., Rosegrant, M. W. & Meinzen-Dick, R. (1997) *Water Allocation Mechanisms: Principles and Examples* (Washington DC: World Bank).

Gibbons, D. C. (1986) *The Economic Value of Water* (Washington DC: Resources for the Future).

Marino, M. & Kemper, K. E. (Eds) (1999) Institutional frameworks in successful water markets—Brazil, Spain, and Colorado, USA Technical Paper No. 427 (Washington DC: World Bank).

OECD (1998) *Water Management: Performance and Challenges in OECD Countries* (Paris: Organization for Economic Co-operation and Development).

Simpson, L. & Ringskog, K. (1997) *Water Markets in the Americas* (Washington DC: World Bank).

World Bank (1993) Water resources management: a World Bank policy paper, (Washington DC: World Bank).

World Bank (2001) *Bridging Troubled Waters—Assessing the Water Resources Strategy Since 1993* (Washington DC: Operations Evaluation Department, World Bank).

World Bank (2002) *Water Resources Sector Strategy—Strategic Directions for World Bank Engagement*, Draft for Discussion of March 2002 (Washington DC: World Bank).

World Water Vision (2000) *A Water Secure World: Vision for Water, Life, and the Environment*, Commission Report (Paris: World Water Council).

Water Pricing in Spain

ANTONIO EMBID-IRUJO
University of Zaragoza, Zaragoza, Spain

Water and Its Economic–Financial System. The Plural Nature of Its Regulations, Techniques and Situations

Explaining the economic–financial system with regard to water in Spain is not the same as talking about the economic–financial system with regard to the Water Act (Embid-Irujo, 1996). Although it may seem paradoxical, the Water Act includes only a fraction—and, it must be said, not the most important—of the economic aspects pertaining to water. In addition to the Water Act, State and Autonomous Community (CCAA) legislation should be taken into account. Table 1 shows the complexity of the problem faced when attempting to acquire an overall awareness of the subject from the standpoint of the legal aspects of the economic system governing water in Spain. Table 1 presupposes the existence of certain forms of payment for water of a certain legal nature received by a public authority, having a certain object and regulated under the terms of a specific law.

Table 1, although extensive, is not exhaustive. For example, it does not include spillway dam charges relating to hydro-electric schemes, which are 'contractual' (included in the concession for the holder of hydro-electric rights), and therefore not taxes. In any event, as can easily be seen, the only complete concordance is in the nature, whatever the receptor, object or legal source. This is not the place to examine in detail the question of whether the specific category of taxation is a Rate (although in most cases it is), or a Tax or a Rate with a mixture of Special Contributions.[1]

Furthermore, this phenomenon of dispersion, variety and singularity that has been indicated becomes even more pronounced if it is considered that the various inter-basin transfers have no semblance of a common economic scheme governing them. This can easily be seen when the various individual water transfer Acts are compared with the regulation of the transfer of water from the Ebro to basins in the Mediterranean Arch as envisaged in Act 10/2001 of 5 July, concerning the National Hydrological Plan. In any event, it is quite clear that there are no elements of 'negotiation' with regard to the determination of the 'price' (which it clearly is not) of water: in all cases, there is a binding determination imposed by the Public Authorities (even though there might have been prior

Table 1.

Act	Object	Public Authority	Name	Nature
The Water Act	Construction of Reservoirs	State	Regulation Charge	Taxation
The Water Act	Other Infrastructures	State	Usage Rates	Taxation
The Water Act	Environmental Care	State	Outflow Control Charge	Taxation
CCAA Act	Infrastructures	CCAA	Various	Taxation
CCAA Act	Treatment	CCAA	Various	Taxation
Local Tax Act	City Water Supply	Town and City Councils	Fee	Taxation
State Laws	Water Transfers	State	Rate-Charge	Taxation
PHN Act	Ebro Transfer	State	Transfer Charge	Taxation

Sources: CCAA, Comunidades Autónomas (Autonomous Communities).
PHN, Plan Hidrológico Nacional (National Hydrological Plan).

consultation with the interested parties, and subject to the presentation of relevant appeals) which ultimately takes on the status of law.

In addition to the above:

(1) When analysing this question the laws to be considered are very diverse: the Water Act,[2] the various Autonomous Community Acts,[3] the Local Taxes Act (1988), the state laws pertaining to the various inter-basin transfers in existence,[4] and finally, the transfer from the Ebro to Mediterranean basins, the National Hydrological Plan Act.[5]

(2) In all cases, it is a public authority which receives the tax. In fact, it can be said that all public authorities of a 'territorial' nature (those who bring together their general, not specific, powers to act, such as institutional bodies in the same territory), have powers to impose taxes on water use, even when this is from different perspectives and objectives: the state, Autonomous Communities and City Councils (it would be better to say Local Bodies, to cover the circumstance in which water is supplied to different Municipalities by municipal Districts or water supply consortia, or even Districts in areas where these local bodies exist and hold powers to act in this matter).

(3) The objective of the taxation is manifold: increased water supply to contribute to the payment of state investment in reservoirs (regulation charge) or other hydraulic infrastructures (usage rates), situations that present similarities between state legislation and that of CCAA with their own basin management (Catalonia, the Canary Islands, the Balearic Islands, Galicia); or environmental care (outflow control charges in the Water Act and something similar in certain CCAA with their own basins), or payment for treatment (the treatment charge imposed by many CCAA), or the payment of infrastructures for water transfers between areas with different Basin Plans, or finally, domestic water supply (rates set by local bodies). In spite of the importance of the above aspect, this paper will not analyse issues such as services granted to private companies or provided by municipal companies, in which the variety of situations and possibilities can be even greater. The complexity that goes with this, especially in the case of urban consumption, is in some cases marked, and the water bills of some Spanish cities (those of Barcelona are typical in this respect) include widely varying charging bases that make it difficult for the consumer to appreciate the reality of the situation.

(4) In no case is the sub-stratum of these figures formed by a 'price' for water as a resource. It is a contribution (which may pretend to be the total) from the beneficiaries of the construction of infrastructures to the public authorities (normally the state, except in the case of CCAA with their own basins, and City Councils) for the investment that has been made.

With regard to the last idea, nobody with broad knowledge of the reality of the field of hydrology would ever completely accept that through these widely diverse types of taxation one was paying the actual cost of water services or compensating for the adverse environmental effects in certain eco-systems resulting from water use. A superficial examination of the widely diverse norms might lead to this conclusion at first, since it is not easy to find references in these texts to say that this is not actually the case,[6] although

Table 2. Average prices of urban supply of drinking water by Autonomous Community

Autonomous Community	Total price (ptas./m³)	Price distribution (%)				Average prices (ptas./m³)		
		Storage + Treat.	Distr.	Treat.	Storage + Treat.	Distr.	Treat.	Storage + Treat. + Dist.
Andalucía	258	14.6	44.2	41.2	38	114	152	106
Aragón	135	15.0	45.6	39.4	20	62	82	53
Asturias	135	18.7	51.6	29.7	25	70	95	40
Baleares	289	17.1	28.8	54.1	49	83	133	156
Canarias	406	18.0	51.8	30.2	73	210	283	123
Cantabra	150	25.0	47.0	28.0	38	71	108	42
Castilla y León	88	22.5	62.3	15.2	20	55	75	13
Castilla-La Mancha	176	20.6	45.5	33.9	36	80	116	60
Cataluña	317	19.9	56.3	23.8	63	178	242	75
Comunidad Valenciana	283	18.1	52.0	29.9	51	147	198	85
Extremadura	195	21.3	38.1	40.6	42	74	116	79
Galicia	108	13.3	45.9	40.8	14	50	64	44
Madrid	227	10.3	48.1	41.6	23	109	133	94
Murcia	362	20.2	40.3	39.5	73	146	219	143
Navarra	129	14.3	48.3	37.4	18	62	81	48
País Vasco	173	26.1	32.2	41.7	45	56	101	72
Rioja	113	22.2	35.8	42.0	25	40	66	47
Ceuta y Melilla	323	13.1	55.1	31.8	42	178	220	103
Total Spain	229	17.3	48.9	33.8	40	112	152	77

Note: 166.386 ptas = €1. Dist. = distribution; Treat. = treatment.
Source: Economic Analyses, National Hydrological Plan, Ministry of the Environment, Madrid (2000, p. 145).

usually, in practice, references to subsidies can be found[7] which in some cases are even included in the legal norms themselves.[8]

A different matter is the actual economic–financial framework in practice, which varies from area to area, and even has certain structural shortcomings. The degree of compliance with the payment of the regulation charge and water tariffs is known to be much higher in the Ebro Catchment Area (for example) than in other areas. Furthermore, there is constant legal wrangling, with attempts in many cases to establish the setting of specific tariffs in order to reduce, if possible, the corresponding payments.

All this means that there is no single concept of water 'price' in Spain, not even in situations that are theoretically identical, since it is possible that there is no similarity, either in practice or in terms of quantity, between one regulation charge for the use of a reservoir and another, depending on the area in which it is situated, when it was constructed or the uses of water. In general, with irrigation water, the 'prices' resulting from the application of regulation charges tend to be on the low side. With the exception of certain cases of infrastructure maintenance, such as the Tagus-Segura aqueduct or of certain City Councils who pass the total cost onto the end-users, these prices may even be symbolic for certain uses, although for subsistence or merely inefficient agriculture, this possibly means costs that are at the limit of economic sustainability.[9] Tables 2–4 clearly show price disparities between regions, cities, irrigation areas, etc. (for these Tables the exchange rates are: €1 = 166.386 ptas.; $1 = 180 ptas. approximately).

Finally, as an exception to the above, there should be mention of the great future potential of the phenomenon of state companies involved in the construction and exploitation of hydro schemes, set up from 1997 onwards, which charge fees (private prices) to their users, and which normally include 50% of the investment costs and the entire exploitation costs.

Table 3. Water prices in different Spanish cities

City and price Ptas/m³		City and price Ptas/m³		City and price Ptas/m³		City and price Ptas/m³	
Barcelona	211	Gerona	102	Lugo	72	Ciudad Real	55
Las Palmas	204	Bilbao	99	Lérida	72	Pontevedra	55
Murcia	191	Castellón	97	Albacete	71	Palencia	54
Alicante	132	Huelva	93	San Sebastián	71	La Coruña	53
Córdoba	127	Oviedo	92	Santander	71	León	50
Madrid	122	Pamplona	91	Orense	70	Segovia	48
Palma de Mallorca	120	Badajoz	87	Salamanca	69	Jaén	39
Almería	119	Zamora	87	Logroño	66	Toledo	36
Cáceres	116	Zaragoza	86	Vitoria	61	Huesca	35
Ceuta	116	Guadalajara	80	Valladolid	61	Melilla	15
Valencia	114	Soria	76	Burgos	60		
Sevilla	112	Málaga	76	Avila	60		
Tarragona	107	Cádiz	74	Granada	58		

Note: The prices are given in ptas/m³ (166.386 ptas = €1).
Source: Economic Analyses, National Hydrological Plan, Ministry of the Environment, Madrid (2000), p. 147.

Table 4. Irrigation prices in different areas and according to different authors

	A	B	C	D	E	F	G	H	I	J
Global										
Irrig. Public initiative	2									
Irrig. with underground water	5–10									
Irrig. with water transfer (Tagus-Segura)	23									
Irrigation in California	2–6									
Guadiana										
Huelva (strawberries)						10				
Mancha Occidental and C. Montiel										15
Guadalquivir										
Genil-Cabra		8								
Fuente Palmera		14								
Fuente Palmera (1990–1991)							9			
Bembézar M.I. (1990–1992)							2–3			
El Viar		2								
Bajo Guadalquivir S.B- XII		5								
Provincia de Granada (average price)									3	
South										
Campo de Dalías				15						
Almería (for red peppers)						30				
Provincia de Almería (ave. price)									7	
Costa de Granada									28	
Segura										
Riegos de Levante M.I.		22								
Campo de Cartagena (groundwater)					25–40	50				
Alto Guadalentín (groundwater)					35–40					
Huerta de Murcia (groundwater extraction)					3–6					
Regadío de Lorca (tariff)					22					
Regadío de Mula (tariff)					8–16					
Mazarrón (groundwater extraction)					25–30					
Aguilas (sale underground water)					60–90					
Mazarrón-Aguilas (groundwater)						50–60				

Table 4. *Continued*

	A	B	C	D	E	F	G	H	I	J
Vega Baja del Segura (for artichokes)										
Provincia de Murcia (ave. price)						20–25			10	
Provincia de Alicante (ave. price)									11	
Júcar										
Acequia Real del Júcar	6	1–3								
Canal cota 220. Onda.		16								
C.U. de Novelda		28								
Vall de Uxó.		25	49							
Bajo Maestrazago									60	
Campo de Liria									28	
Cenia-Maestrazgo										
Mijares-Plana de Castellón								10–19		9–87
Palancia-Los Valles								6–29		
Turia								15–27		
Alarcón-Contreras								9–11		
La Safor								1–21		
Serpis										1–23
Marina Alta									7–57	
Marina Baja										6–85
Vinalopó-Alacantí-Vega Baja									12–21	
Alto Vinalopó									18–65	
C.R. Novelda (1989)										5–36
SAT en el Vinalopó										31
Provincia de Valencia (ave. price)									5	3–40
Provincia de Castellón (ave. price)									7	
Average price Comm. of Valencia								19		

Notes: The price is in ptas/m^3 (166.386 ptas = €1).

Sources: Works from which these data were extracted according to the original source: Economic Analyses, National Hydrological Plan, Ministry of the Environment, Madrid (2000), p. 149; (A) Sumpsi Viñas *et al.* (1999), p. 149; (C) Avellá, in: Sumpsi Viñas *et al.* (1999), p. 149; (D) Naredo *et al.*, in: Sumpsi Viñas *et al.* (1998), p. 149; (E) Albacete & Peña (1995); (F) Morales Gil (1997); (G) Losada y Roldán (1998); (H) Carles *et al.* (1998); (I) MOPTMA; (J) Caballer & Guadalajara (1998).

Economic Criticism of Traditional Management: The Water Market and the Real Cost of the Resource

One of the results of the complex phenomenon outlined in the previous section is that of the lack of stimuli for proper water management, especially for irrigation. Very low water prices are not conducive to saving, and besides, in many cases, such saving would be extremely difficult to achieve in view of the poor state of the supply lines which generate losses.[10] The result of this inefficiency is often the constant desire to build new hydro schemes to offset already existing management problems. But at the same time, the system is not capable of generating economic resources to stimulate infrastructure improvement procedures, and 'modernization' of irrigation resources is a recurrent theme, even when bereft of substantial practical actions, at least up to now.[11]

In view of the above, in the late 1980s and 1990s in Spain and other similar countries, a substantial economic criticism was formulated, with some legal basis,[12] of this type of management. Solutions were found to the problems that currently existed with the introduction of market techniques (transactions between individuals, the purchase of rights by the Public Authorities) and the subsequent setting of a 'real' price for water which would be paid by users or beneficiaries of the infrastructures. For certain economists or the intellectual colleagues of certain economists, this policy was a sort of 'magic wand' that would solve all the current problems at a stroke, while other experts were more realistic, even though they had set out from the same starting point, and they envisaged this policy within the framework of other actions which would necessarily accompany them (Perez-Diaz *et al.*, 1996).

There can be no question of the professional and intellectual integrity of the vast majority of the contributions in this field, made by accredited experts. Furthermore, it is difficult to argue against the sheer logic of numbers and their consequences, even though they often ignore an extremely important element when talking about water: the sociological aspect, the attitude of certain users in this respect and the existence of myths and creeds which are often much more firmly held than may have been realized. Similarly, this doctrine does not fully take into account the importance, from a regional planning point of view, of certain agricultural usages which are probably not entirely justified in economic terms but perfectly justified in others, referring in 'sociological' terms, of course, to genuine direct growers, people who wish to live in, and off, land that can be irrigated. Thus, there is no reference here to the numerous economic interests, generally not 'territorial' in nature, which in recent years have been found in irrigation agriculture based on pure profit (or similar criteria) and quite apart from any other consideration.

The Theoretical Establishment of a 'Water Market' and Why It Is Not Possible in Practice. Cost Recovery in Water Management as per European Directive 2000/60/EC

The doctrine under consideration has clear effects in Spain and directly determines some of the content of Act 46/1999 of 13 December, concerning the reform of the Water Act (today included in articles 67–72 of the revised Text of the 2001 Water Act). Specifically, this text regulates what is called a 'transfer of water usage rights contract', and envisages, under the auspices of Catchment Boards in certain circumstances (related to drought), the establishment of Water Usage Rights Exchange Centres, taking as their reference the California Water Bank. The proclaimed objective is to improve management practices with existing resources and this would favour the environment in that it would avoid the construction of certain hydro works.[13] Water would, in line with the economic premises envisaged, unfailingly seek, through market mechanisms, the most profitable uses and would leave aside any usages that were less profitable and environmentally more harmful. At this point, it should be remembered there was a great deal of political discourse,[14] scientific conferences and information that generally revolved around this question in the late 1990s.[15]

However, these hopes and controversies have not accompanied the development of what is today articles 67 *et seq.* of the Revised Text of the Water Act. Over two years after these innovations in the regulations, it can be seen that there has effectively been no development from the novelties of Act 46/1999. Consequently it seems hard to enter with any security or legitimacy into a water usage rights transfer contract, because fundamental questions such as the volume of rights to be transferred or the transfer price are deferred by the Act to the Regulations, and these have not appeared yet (draft regulations have been published but the date of approval is as yet unknown).

This is obviously a pity. Notwithstanding the legal shortcomings, at least, as far as the author is concerned, in public control of the transfers (for example, the administrative silence regarding authorizations for contracts, which operate with an extremely short deadline, and is therefore unconstitutional because it breaches the very principles of the legal framework of the public domain), one welcomes and continues to be optimistic about a mechanism that enables the transference of rights which, added to other mechanisms already in place concerning demand management,[16] could well lead to an effective improvement in the current situation under certain circumstances.

With regard to the question of the 'price' of water, current internal debate is now affected by the declaration of the European Parliament and Council Directive 2000/60/EC of 23 October 2000, establishing a community framework for action in the field of water policy: article 9 establishes the principle of the "recovery of the costs of water services". The text of article 9, which has undergone several changes in the various versions of this Directive in its long, drawn-out drafting process (the original project dates from 1997), is highly complex and full of nuances, although it does take a clear line with regard to transferring costs to the beneficiaries of water services, including 'environmental' costs, which places the intellectual principles of the question on completely new ground, quite different to what has been experienced in Spain up to now. The future evolution of this question is certainly problematic, and the immediate short-term effects may perhaps be ruled out, although it would be extremely difficult to argue successfully against the trends that are marked out. Obviously, it is going to take some time to define precisely what these 'environmental costs' are, and also any exceptions to the recovery of costs that may be made. However, these will be exceptions, not general rules. This question seems quite clear in that it is also in line with Western European thinking on the problem of rational use of natural resources, and not just water.

A Prolonged Era of Transition. The Old and the New Water Law. Ambiguities in the National Hydrological Plan

All the above leads to the conviction that Spain is in the midst of a phase of profound conceptual instability with regard to the economic framework of water usage, and it is a phase of instability that extends to the very principles of water usage, in view of the importance and relevance of economic considerations regarding the entire legal framework.[17] There are questions concerning hydro policy contained in the debate that have been summarized earlier, not just questions concerning the economic framework.

In this respect, it should be pointed out that Spanish hydro policy has recently centred on the National Hydrological Plan, which does not contain any decisive elements on this question. On the one hand, it envisages the carrying out of studies "for the gradual implantation of the principal of cost recovery and justified exceptions" (supplementary provision 11), while on the other hand, in connection with the most important infrastructure that it contains and the one that is destined to be the longest-lasting, the Ebro-Mediterranean Basins transfer, it sets up an economic–financial framework (c.f. article 22) based on the existence of subsidies, in as much as the 'usage quota' (which is one of the two components of the transfer quota) is formed by taking into account the 'recoverable' investments, in other words, not the entire investment. Second, it sets up an environmental component (environmental quota) of the transfer charge to be paid by the users. However, this is extremely low ($€ 0.03/m^3$ of transferred water), and will hardly provoke changes in consumption to bring about a more environmentally sustainable usage of natural resources. The problem is that this is the object of all environmental taxes, a generic category in which this environmental quota should be included.

To summarize, these are times of evolution with principles originating from new forms of management, and some of the old ones. A new direction shows the prevalence, albeit not absolute, of economic (and of course, environmental) considerations in the use of water, but a direction being followed along a rocky road in which Spain, because of its specific situation with regard to water resources, has a good deal at stake.

Notes

1. Without going into the legal details, for reason of space, in Spanish Law taxations are divided into the three types specified in the text: Tax, Special Contribution and Rate.
2. cf. the revised text of the Water Act approved by Legislative Decree 1/2001 of 20 July.
3. The most complex and detailed legislation is that of Catalonia, but with regard to water treatment, there is also legislation in Galicia, Asturias, La Rioja, Navarre, Aragon, the Community of Valencia, Murcia, Madrid, the Canary Islands and the Balearic Islands. Other CCAA also have legislation regarding urban water supply: Catalonia, Madrid, Asturias and the Canary Islands.
4. See the legal regulation of the Tagus-Segura transfer (Acts passed in 1971 and 1980 with certain subsequent modifications), the Ebro transfer to Tarragona (1981, with certain subsequent modifications), the Tagus-Guadiana transfer (1995, with certain subsequent modifications), the Guadiaro-Guadalete transfer (1995), and the Guadalquivir-South transfer (1999).
5. cf. Act 10/2001 of 5 July, concerning the National Hydrological Plan.
6. One only has to read the regulation and water tariffs precept in article 114 of the revised text of the Water Act to arrive at the conclusion that the beneficiaries pay the total cost of the state's investment.
7. For example, when certain reservoirs are attributed, in many cases quite rightly so, with the ability to mitigate flooding, implying because it is related to general interests, the assumption of a specific percentage of the cost of construction and exploitation by the public authorities.
8. For example, once again, the reference to the fact that only 'recoverable' investments (and this implies that there would be other investments that are not recoverable, without any indication of which is which) would be taken into account when determining the transfer fee regulated by the National Hydrological Plan. cf. article 22.6.a.
9. cf. Sumpsi Vinas (1998); Lopez-Galvez & Naredo (Eds) (1997); Various authors (1997).
10. This phenomenon is not exclusive to agricultural use; many of the infrastructures for urban supply are also obsolete, leading to losses of 30% of the volume transported, and sometimes more.

[11.] 'Up to now', because the National Irrigation Plan, as yet to be wholly approved at the time of writing, aims to carry out significant actions in this respect.

[12.] For example, the clause regarding rational use of natural resources contained in article 45.2 of the Spanish Constitution, which incidentally concurs with article 174 of the European Community Treaty. For further information, see Embid-Irujo (1997, p. 209).

[13.] This policy is accompanied by noteworthy efforts in the field of the regulation of wastewater recycling and desalination. cf. Embid-Irujo (2000).

[14.] Corresponding with the setting up in 1996 of the Ministry of the Environment, which logically envisaged water resources as essentially a natural resource, whose premises were certainly not the construction of hydro works as practically the only aspect of hydrological policy, as had predominantly been the case in the past. A good example of what is being said can be found in the Green Paper on Water (1998). cf. Embid-Irujo (1999).

[15.] Much has been written on the regulation contained in articles 67–72 of the revised Text of the Water Act, but the work by Vazquez (2000) is particularly recommended.

[16.] On several occasions the author has expressed objections to custom-built recipes for improving existing problems. It is much better to have several so that, depending on the circumstances, the most suitable option can be chosen at any given moment. Water usage rights transfer contracts can solve certain problems, although they obviously cannot be considered a universal panacea (Embid-Irujo, 2001).

[17.] References to these questions can be found in Embid-Irujo (2001, p. 59).

References

Various authors (1997) *La Economía del Agua en España* (Madrid: Fundación Argentaria).

Albacete, M. & Peña, M. (1995) Consideraciones sobre algunos aspectos económicos de la ordenación y las disponibilidades de agua para riego, in: M. Senent & F. Cabezas (Eds) *Agua y Futuro en la Región de Murcia* (Murcia: Asamblea Regional de Murcia).

Caballer, V. & Guadalajara, N. (1998) *Valoración Económica del Agua de Riego* (Madrid: Mundiprensa).

Carles, J. *et al.* (1998) Precios, costos y uso del agua en el regadío mediterráneo, *Congreso Ibérico Sobre Gestión y Planificación de Aguas*, pp. 231–256, Zaragoza.

Embid-Irujo, A. (1996) Condicionamientos jurídicos de una política de precios del agua, in: A. Embid-Irujo (Ed.) *Precios y Mercados del Agua* (Madrid: Civitas).

Embid-Irujo, A. (1997) La utilización racional de las aguas y los abastecimientos urbanos. Algunas reflexiones, *Revista Aragonesa del Agua*, 10, pp. 209–219.

Embid-Irujo, A. (1999) *Planificación Hidrológica y Política Hidráulica (El Libro Blanco del Agua)* (Madrid: Civitas).

Embid-Irujo, A. (2000) Reutilización y desalación de aguas. Aspectos jurídicos, en las págs. 113 y ss. de, *La Reforma de la Ley de Aguas. (Ley 46/1999, de 13 de diciembre)* (Civitas: Madrid).

Embid-Irujo, A. (2001) Evolución del derecho y de la política del agua en España, *Revista de Administración Pública*, 156, pp. 59–100.

Lopez-Galvez, J. & Naredo, J. M. (Eds) (1997) *La Gestión del Agua de Riego* (Madrid: Fundación Argentaria).

MOPTMA (1995) *Medidas y actuaciones de racionalización del uso de la oferta del agua y su incidencia en balances hidráulicos en las cuencas del Sur, Segura, Júcar e Internas de Cataluña*, informe intgerno, Madrid.

Morales Gil, A. (1997) *Aspectos Geográficos de la Agricultura de Ciclo Manipulado en España* (Alicante: Universidad de Alicante).

Perez-Diaz, V., Mezo, J. & Alvarez-Miranda, B. (1996) *Política y Economía del Agua en España* (Madrid: Círculo de Empresarios).

Sumpsi Vinas, J. M., Garrido Colmenero, A., Blanco Fonseca, M., Varela Ortega, C. & Iglesias Martínez, E. (1998) *Economía y Política de Gestión del Agua en la Agricultura* (Madrid: Mundi-Prensa y Ministerio de Agricultura, Pesca y Alimentación).

Vazquez, C. (2000) La regulación de los contratos de cesión de derechos de uso de agua, in: A. Embid-Irujo (Ed.) *La Reforma de la Ley de Aguas*, p. 159 (Madrid: Civitas).

The Functions, Impacts and Effectiveness of Water Pricing: Evidence from the United States and Canada

CHARLES W. HOWE

Institute of Behavioral Science, University of Colorado, Boulder, CO, USA

Introduction

There are many kinds of the commodity that we call water, differentiated by location, method of delivery, extent of treatment, reliability and other dimensions of quality. The costs of bringing these supplies to their users differ greatly, so it can be expected the prices charged to the users also differ. As an example of the complexity of what is meant by 'the price of water', consider the situation in a well-known water district in northeastern Colorado, The Northern Colorado Water Conservancy District. This District distributes about 308 million m^3 of raw, untreated but high quality water per year to irrigators, rural water districts, large industries and towns in northeastern Colorado (see Tyler, 1992 for a detailed history). The District charges farmers a very low price of about $6 per 1000 m^3 for water delivered to their farms. Cities and industry pay about $16 per 1000 m^3. These prices are administratively set by the District. The costs of the District that are not covered by these low prices are covered by the collection of real estate taxes levied against all urban and rural real estate in the District.

In addition to the low prices and real estate taxes just mentioned, two types of market transactions are found in the Northern District: permanent sales of water and one year 'rentals' of water. The current market price for the permanent sale of water is about $9000 for a permanent annual supply of 1000 m^3 or about $450 per 1000 m^3 per year if a 5% interest rate is used. The market for the temporary use water (the 'rental market') allows towns that own excess water supplies for drought protection or in anticipation of growth to 'rent' water to other agriculture on a one-year basis. Towns frequently provide such water cheaply to maintain a rural, open space environment close to the town. In years of average rainfall, these 'rental' prices range from $18 per 1000 m^3 to $36 per 1000 m^3. If rentals occur later in the growing season, prices are usually higher but rentals are usually arranged in the early spring season so that farmers will be able to anticipate their water supplies. In drought years, rental prices can be much higher or water rentals may cease, as is the case with severe drought in the western US.

The towns that use bulk water from the District use various pricing methods. Some small towns still do not meter residential users, so that customers pay a monthly fee that does not depend on the volume of water used. Most towns, however, are metered and increasingly are using an increasing block rate structure. In addition, there is usually a wastewater treatment fee that depends on sanitary sewer flow (estimated by winter water use) and a flood control fee that depends on property location.

So, what is the price of water in northeastern Colorado? There are different 'prices', depending on the type of water service being provided, the decision being made, the revenue structure, or whether or not a water market is accessible to the water user. The correct definition of 'water price' should be: the charge or market price that would affect a rational water user's decisions concerning their pattern of water use, including quantities of water and water-related investments. This is the 'behaviourally relevant price', i.e. the cost that a 'rational water user' will compare with their 'marginal benefits' in deciding on a water use technology and water quantities. The farmer applying irrigation water to low value crops within the Northern District will disregard the very low price charged by the District and will immediately understand that the (much higher) market price for water is the relevant 'price' or 'opportunity cost' for making water decisions. A homeowner facing an inefficient price structure that makes the monthly charge a function of irrigated garden area will take that charge into account when designing the garden, but not when deciding on the volume of water to apply to the resultant area.

The Conflicting Roles of Water Price

Among the primary roles of water price are: (1) the 'economically efficient' allocation of existing supplies in the short term; (2) the generation of adequate revenues for the operation, maintenance and expansion of the water system; and (3) the 'equitable' treatment of water users. In the short term, the delivery capacity and /or raw water supply of a supply system is fixed. Since price affects the quantities users demand, price can be used by water managers to adjust demand to the available supply. During drought periods when demands are high and supplies low, price can be raised to equate supplies and demands. However, short-term price changes are often administratively or even politically difficult.

Why is price an 'economically efficient' way of rationing fixed supplies of water among users? The economic efficiency argument for price is that it might be assumed that water users compare their marginal benefits with the marginal costs they face when deciding on changes in water use patterns. If all water users of a particular class (such as single family residential users) face the same price, each will adjust their water use until their marginal benefits fall or rise to equality with the price. All users in that class then exhibit the same marginal benefits, a common sense condition for maximizing the *total* benefits from the use of available water. Naturally, there are other ways of rationing the available supply such as requesting cutbacks, requiring even-odd day garden watering or prohibiting certain uses. While these rationing mechanisms may work (see later section on conservation effectiveness), none of them guarantees that water will be distributed to the users who place the greatest value on the water, nor can users choose what types of use are more valuable to them.

The second major function of price is the production of revenues for the water supply agency. Naturally, all components of the 'revenue package' contribute to revenues. Whether increases in price (the volumetric charge) will produce more or less revenue depends upon the responsiveness of users to price, i.e. the 'price elasticity' of the users' demand for water. There is a vast literature on the price elasticity of water demands in residential, industrial and agricultural uses (see Foster & Beattie, 1979; Billings & Agthe, 1980; Howe, 1982; Boland *et al.*, 1984; Schneider & Whitlatch, 1989; Nieswiadomy, 1990; Tate, 1990; Renzetti, 1992). The consensus is that residential and industrial demands (except for cooling water) are 'inelastic' while agricultural demands are 'elastic.' For 'inelastic' demands, when price is raised, revenue collections also will rise; while for 'elastic' demands, an increase in price will reduce revenues. An important issue is the stability of the revenues generated by the price structure (see later section on revenue stability).

A third consideration in water pricing is fairness or equity among water users. It is often asserted that raising the price of water will 'hurt the poor' or be unfair to them. This is a major issue in Third World countries where much of the population is at subsistence level. The public health values of low quantities of daily water taken from a common standpipe are sufficiently high and the awkwardness of trying to collect fees for small quantities sufficiently great that a zero price may be justified. In countries where all residences are metered, the fairness issue can be handled through 'life line pricing', e.g. with a very low first block in the rate structure. This is not to say that poverty is not a problem in advanced countries but that lowering the price of water for all users is an inefficient way of dealing with the problem.

Water Pricing in the United States

The Provision of Bulk Water Supplies

Bulk water supplies (i.e. untreated water from an original source) are obtained in the US either from the private appropriation of surface and groundwater supplies through private infrastructure or from larger projects undertaken by individual states or the Federal government. Early water development was entirely private, both for agriculture and for town uses. State and Federal governments got involved in water provision as further water source development became more costly and technologically more complex. Both private and public water developments must abide by existing state regulations, permitting systems and/or the rules for acquiring water rights.

Water supply abstraction or development in the eastern US continues to be governed largely by the English 'appropriations doctrine' of water law that evolved from Roman 'prior occupancy' law in response to the changing technologies of the Industrial Revolution that initially were based on water power, i.e. water mills that consumed very little water and usually had little effect on water quality. 'Appropriations doctrine' allows all riparian land owners 'reasonable use' of water from rivers, ponds and lakes, provided they return water to the stream 'unimpaired in quantity and quality', a condition obviously impossible to achieve, but not grossly inappropriate in water-plentiful regions where most uses are non-consumptive. It is not appropriate for water short regions where the major uses are highly consumptive. Some of the eastern states are now converting to permit systems in recognition of increased pressure on water supplies.

As the much drier western US developed, irrigation was necessary for agriculture and, at times, the better lands were at a distance from the rivers. Thus a new legal doctrine of 'prior appropriation' evolved that allowed the establishment of a 'water right' to be used on either riparian or non-riparian lands, that gave a priority (preference) according to the date of first use and that allowed significant consumption of the water diverted from the source (the consumptive fraction becomes part of the definition of the water right). In the 19th century, these 'priority water rights' were declared by the courts to be personal property that could be bought and sold separate from the lands where they were originally used (Scott & Coustalin, 1995).

While permits are increasingly required in the eastern states and while water rights must be acquired in the western states (by establishing new rights or purchasing existing ones) in order to use water, there are no significant charges for the abstraction of water from the source. The eastern state permit systems increasingly are allowing trading. Since water rights are tradable in all western states, markets for permits and rights increasingly generate the relevant prices for water.

The Provision of Irrigation Water for Agriculture

Irrigation is not common in the eastern US, although it is increasingly used as a supplemental water supply during drought. In the western parts of the country, irrigation is necessary for many crops, although most wheat and some cotton are grown without irrigation. Early development (say before 1900) was by groups of farmers who formed ditch companies to built common diversion, storage and distribution works. Again, there were no charges for taking the water from the stream, and charges by the ditch company to member farmers just covered operating costs and debt repayment. These practices continue now, although active markets have developed for ditch company shares, so farmers are quite aware of the opportunity cost of their water. A vote of the membership is usually required before a member can sell his/her water.

Irrigation supply developments after 1900 were mostly by agencies of the federal and state governments, especially the Bureau of Reclamation that was established in 1902 to assist local groups in the financing of projects. A revolving fund was established for 10-year loans, the repayment of which would allow other projects to be funded. The saga of the Reclamation Program has been one of inability to repay capital costs, stretching out the repayment period and finding other ways of reducing the real value of farmer repayments. Currently, the repayment period for a full supply project (as opposed to a supplemental supply project) is 50 years, with no repayment for 10 years, no interest on the unpaid balance and no allowance for inflation. Repayment of capital costs averages around 15% in real terms. Operating costs are covered by charges to the farmer based on acreage and by real estate taxes that bear no relation to water use. The conclusion is that agricultural water pricing is badly carried out in the US and results in great inefficiency in agricultural water use and in the allocation of water between agriculture and other sectors.

Markets for agricultural water do exist. Although imperfect, they help to overcome the inefficiencies of typical irrigation water charges by confronting the farmers with a much more realistic estimate of the opportunity cost of the water (see Howe & Goemans, forthcoming).

The Provision of Municipal and Industrial Water

The structure of the municipal water supply and wastewater sector. Cities in the US typically develop their own surface and/or groundwater supplies, but in some cases it is possible for them to contract for raw water from state or federal projects. Historically, town water supplies were developed and marketed by private providers. As cities grew, provision of water became a major function of local government for both technical and political reasons. This trend toward public sector provision accelerated in the 1920s when the federal Congress authorized tax-free municipal bonds, giving municipalities a financial advantage.

The urban water industry is highly diversified today with nearly 54 000 community systems. However, 85% of the systems serve only 10% of urban populations (National Research Council, 2002, chapter 1). Roughly 40% of the systems supply treated water only, with half also providing wastewater treatment. Other services also provided include solid waste, energy, irrigation water, storm-water control and reclaimed water, as indicated by the diagram below from a survey by the American Water Works Association (AWWA) for 1992.

Private utilities (investor-owned) account for about 15% of urban supplies and a much smaller part of wastewater utilities. Privately owned utilities are regulated by state utility commissions, while publicly owned utilities are not. However, operation of publicly owned systems through contracts to private firms has grown rapidly in recent years (National Research Council, 2002). About 65% of the utilities use surface water, 25% use groundwater and the rest buy treated supplies from other utilities.

Pricing by urban water utilities. A survey in 1992 by the AWWA (1992) found the following characteristics for residential water use:

- Average residential water withdrawal per year: 415 m^3 (110 000 gals)
- Average population/connection: 4.1 persons
- Average residential production per capita per day: 0.679 m^3 (180 gals; regional range: 150 to 217 gals)
- Average residential water cost per year (1992 $): $182

A 1996 AWWA survey showed the following distribution of rate structures for the residential sector: uniform rate 39%; declining block rate 33%; increasing block rate 22%; seasonal 2%; and flat rate 4%. It is surprising that 33% of the responding utilities still use declining block rate structures, although this percentage has decreased from 39.5% in the 1992 survey. Uniform rates (constant price per unit volume) have decreased from 45.6%, while increasing block rates have increased from 14.7% to the 1996 level of 22%. The steepness of increasing block structures has also increased in the last decade. 'Flat rate' in this diagram refers to a fixed monthly charge only, while 'seasonal' refers to uniform rates that rise in the hot season.

In the early 1990s in response to sustained drought, the Los Angeles Department of Water and Power instituted a residential 2 block system that varies from climatically normal years (low block of $1.71, high block of $ 2.92) to shortage years (10% shortage: low block $1.71, high block $3.70; 25% shortage: low block $1.71, high block $6.05). (Hall & Hanemann, 1996). Boulder, Colorado has a monthly 3 block structure: low block up to the account's average monthly winter 'consumption' (AWC)

@ \$1.60 per thousand gallons; second block for withdrawals greater then AWC but less than 350% of AWC @ \$2.85 per thousand gallons; high block for over 350% of AWC @ \$4.25 per thousand gallons. In addition, there is an \$8 monthly service charge; a monthly winter (December–March) wastewater fee based on actual use with other months based on AWC or actual use, whichever is lower; and a storm water/flood management fee that depends on location and impermeable area.

Frequency of billing has been shown in several studies to influence residential withdrawal demand, with greater frequency generating greater awareness of the relationship of cost to volume of use (Espey *et al.*, 1997; Howe & Linaweaver, 1967). The AWWA 1992 survey showed that 44% of responding utilities billed each month, 21% every 2 months and 29% every 3 months. With improved meter reading technology and the frequent 'outsourcing' of meter reading, monthly readings should become the standard.

Commercial and industrial users face uniform or even decreasing block structures, the latter perhaps justified by economies of scale for treatment and delivery of large volumes. One can speculate that the decreasing block is sometimes justified by the belief that this will help attract commercial and industrial enterprises.

A serious shortcoming in most US water pricing, both urban and rural, is that the real opportunity cost of the raw water is not counted as a cost and is, therefore, not reflected in the price structure. In the eastern US, no charges are made for the extraction of water from basic sources under the 'appropriations doctrine'. In the western US, only low administrative fees are usually charged for establishing new water rights if they are available. Only if water rights have to be purchased are their purchase prices recognized as costs. Even then, municipal accounting practices frequently fail to account for assets, including water rights, so that once purchased they are not treated as assets of value. In Boulder, Colorado, the water utility has a sophisticated pricing scheme for water, wastewater and flood control as noted earlier in this section. The fact that the water rights that underlie the basic supply have very high market values is not reflected in the prices charged. If they were, water prices would be nearly double their current levels-which they should be from an economic efficiency point of view.

The importance of price and income elasticities of demand. The formulation of appropriate pricing policies for urban water requires knowledge of the price and income elasticities of demand. These elasticities have implications for forecasting demand, for revenue changes from changes in price structures and for the effectiveness of price for achieving conservation and reducing peak demands on the system. The dependence of price responsiveness of demand on the income levels of water users is an important issue, i.e. the cross-elasticity of price with income.

The literature on price elasticities is vast, but a range of values will be quoted here. Howe, in a reworking of data from the 1967 study, found that price elasticity estimates were lower than in the earlier study: a very low value of -0.06 for winter (in house) demands, -0.57 for summer demands in the well watered east and -0.43 for the semi-arid west (Howe, 1982). These estimates imply that pricing could be effective in reducing summer demands but would have essentially no effect on in-house use.

A 1997 study of two towns in California (Renwick & Archibald, 1997) estimated the price elasticity of residential demand at -0.33, while a 1999 study of seven southwestern cities (Michelsen *et al.*, 1999) showed that elasticities were significantly different for each city with all estimates lying below -0.37 and the elasticity across the aggregated sample of -0.12. A study (Renwick & Green, 2000) that included various conservation programme variables in the demand function arrived at an estimate of -0.16. Thus it is clear that location, season and the presence of other conservation programmes affect the responsiveness of demand to price.

Income elasticities have been estimated by several studies. The 1967 Howe & Linaweaver study gave values around 0.30. The Renwick/Archibald study gave an estimate of 0.33, while the JEEM study gave an estimate of 0.25. Thus there seems to be consistency on this parameter, implying that a 100% change in household income would lead to something like a 30% increase in water demand.

As noted above, the cross-elasticity of price with income is important to water planning in a changing income environment. None of the studies directly estimated this cross-elasticity but the importance of income level to price responsiveness of demand is shown by the price elasticities by income level estimated by Renwick & Archibald (see Table 1).

Thus average elasticities will not do the job with a move from low-income towns or neighbourhoods to high-income areas.

The 'peak demand' problem. An important management concern is 'the peak period demand problem' faced by most utilities and that relates closely to pricing. The AWWA 1992 survey showed regional average ratios of peak day to average day deliveries ranging from 1.37 to2.36, but the ratio can be much higher in some places. The major issue here is the adequacy of treatment and delivery capacity, an issue even when the raw water supply is sufficient. Providing capacity to meet all peaks can be costly, so the issue is what price and non-price steps can be effective in reducing peak demands. There is a large literature on peak-load pricing since it faces all distribution systems (see Crew *et al.*, 1995). Other steps include alternate day outside use restrictions, education and information programmes, prohibitions of various outside uses, etc. The key to the usefulness of price is the responsiveness of short-term demands to price under the climatic conditions that typically lead to peak season, peak day or peak hour demands.

Howe & Linaweaver (1967) long ago found significant responsiveness of peak day demands to marginal price, with elasticities ranging from -0.30 in metered areas of the US west to -1.25 in eastern metered areas. This was a surprising result since there was no peak day pricing that would penalize use on a particular day. Refinements in data and estimation techniques today indicate elasticites in the range of -0.3 to -0.20. These elasticities are highly dependent on income levels and on the simultaneous use of other steps along with price (see Michelsen *et al.*, 1999 and Renwick & Archibald, 1997). The point is that with a (conservative) value of -0.2, it would be necessary to raise price 21/2 times to achieve a 50% reduction in maximum day demands—a political and administrative impossibility. Thus non-price steps like alternate day garden watering must be the main instruments for dealing with the peak period demand problem.

Table 1.

Total sample	-0.33
Income $<$ \$20 000	-0.53
\$20,000 $-$ \$59 999	-0.21
\$60,000–\$99 999	-0.22
Income $>$ \$100 000	-0.11

Renwick & Archibald found significant reductions in demand during drought associated with money rebates for low flow toilets and shower heads, prohibitions of various outdoor uses, and a pricing system that severely penalized households exceeding a certain percentage of historical use by the household. In the JEEM study of nine California towns, the following steps were found significant in reducing demand during the 1986–96 period that covered severe, continuous drought through 1992: public information and education programmes; distribution of retrofit kits for showers and toilets; and various rationing policies.

Water Pricing in Canada

The federal government of Canada passed the Canada Water Act in 1970 and created the Department of the Environment in 1971, entrusting the Inland Waters Directorate with providing national leadership for freshwater management. Under the Constitution Act (1867), the provinces are 'owners' of the water resources and have wide responsibilities in their day-to-day management as is the case in the US. The federal government has certain specific responsibilities relating to water, such as fisheries and navigation as well as exercising certain overall responsibilities such as the conduct of external affairs (the above quoted from Environment Canada, 'Water Policies of Canada', available at http://www.ec.gc.ca/water).

While providing national leadership to ensure that Canada freshwater management is in the national interest, Environment Canada also actively promotes a partnership approach among the various levels of government and private sector interests that contribute to and benefit from the wise management and sustainable use of the resource.

All of these interests were extensively consulted during the 1984/85 Inquiry on Federal Water Policy, which conducted Canada-wide hearings toward the development of a federal water policy. Guided by the findings of the Inquiry, the government released its Federal Water Policy in 1987, which has since given focus to the water-related activities of all federal departments and which will continue to provide a framework for action in the coming years as it evolves in the light of new issues and concerns (the above was taken from Environment Canada, 'Federal Water Policy', available at http:// www.ec.gc.ca.).

The percentage of residences that were metered was unchanged between 1996 and 1999 at 56%, a small increase from earlier years. In 1999, 43% of households with municipal water service paid flat fees. Some 3.4 million households presumably on a volume-based system fell within the basic allowance covered by the fixed monthly fee, so were essentially on a flat fee basis.

The constant unit charge was the volume-based charge most frequently used, with 39% of ratepayers facing this structure in 1999. The average unit charge was (Canadian) $0.96. Declining block rates accounted for 13% of ratepayers, down sharply from earlier years. Increasing block rates or more complex structures accounted for 9.9%, up from the early 1990s. The first block averaged $1.12, up slightly in real terms from earlier years. The average monthly charge at the 25 m^3 level of use was $29, although charges varied widely over the provinces, from $15 in Quebec to $87 in the Northern Territories.

Charges for wastewater collection and treatment are usually integrated with water charges. Flat fees are most frequent, followed by percentages of the water charge often running from 50% to 100% of the water charge. Overall, sewer charges accounted for 39.4% of the average total bill.

The trend in per capita water withdrawals had been down, to 327 litres in 1996 but jumped to 343 litres in 1999. Flat rate consumers showed much higher rates of use than those on volumetric structures.

Commercial water use increased to 2.84 million m^3 per day in 1999 from 2.74 million in 1996. Commercial rates rose more slowly than residential rates up to 1999, the national average charge per service being $37, ranging from $22 in Quebec to $58 in Alberta. Overall, residential daily use accounted for 53% of the 1999 total rate, commercial use 20%, industrial use 14% and 'other' 13% (data taken from Environment Canada, 'Municipal Water Pricing, 1991–1999' available at http://www.ec.gc.ca/erad/dwnld/waterreportbw.pdf).

The Relevance of Pricing to Instream Water Quality

In the administration of ambient water quality, appropriate water pricing can again play an important role (the reader is referred to the excellent Inter-American Development Bank Study, *Investing in Water Quality, 2001*). Imagine a town that uses water and returns a fraction of it to the municipal wastewater treatment plant. The treatment plant, after primary, secondary and perhaps tertiary treatment returns the flow to the river. How is pricing relevant to this cycle of water use and waste disposal?

First, prices charged residential, commercial and industrial users will affect the quantity of water they withdraw and thereby the volume that returns through the sanitary sewer for treatment. The concentration of pollutants in the waste flow is thus affected and will affect the costs of treatment. While residential users have little freedom in adjusting their water-borne waste loads, commercial and industrial water users have a wide range of choice over technology, inputs and products that affect their water-borne waste loads. Thus, charges on the waste flows from commercial and industrial establishments (i.e. prices for pollutants) will induce changes to lighten the pollution load.[1]

In the final step of this chain of events, the treated effluent is poured into the river, and the sanitized sludge is applied to farmland. The latter disposal operation is beneficial to farmers but involves costs of sludge treatment, transport and application. The nutrients that finally enter the river from the treatment plant cause damages to other water users, and these damages should be reflected in charges (taxes) by the water quality control authority against the outflow of nutrients. These discharge taxes are widely used in Europe but not in the US.

Pricing and the Importance of Service Reliability

A forthcoming US National Research Council report (National Research Council, 2002) on water utilities gathered information that showed that most urban water customers in the US are willing to pay more than historical rates for higher reliability of treated water supplies. The failure of supply to meet demands during droughts leads to water user losses, usually not very large, but in extreme situations can lead to large out-of-pocket losses for the replacement of gardens, lawn, trees. In the 1976–77 drought in the western US, it was estimated that single family homes in Marin County, California averaged $570 for the replacement of lost plants. Replacement costs are usually far below what customers would be willing to pay to avoid damage to their lawns and gardens. In cases of complete failure of supply, common in some countries, the dangers of contamination and related health risks become important.

Public officials frequently face trade-offs between risk and cost: the safety of highways, airlines, safety of the work place and water supply reliability. Urban water utilities are beginning to use explicit risk-based decision procedures for both long-term planning and short-term shortage management but many still use procedures based on 'the drought of record' or similar events. The key question is whether or not water users are willing to pay enough to cover the costs of reducing the frequency of shortage events. Techniques have been developed to measure water users' willingness to pay for increased reliability or their 'willingness to accept' compensation for decreased reliability, the latter being the case in Boulder, Colorado that has an overly reliable system (Howe & Smith, 1993; Howe, 1994). The point is that the level of reliability should be explicitly optimized and the costs reflected in the price structure.

Concluding Remarks

Water prices that are appropriately set and applied at different points of the water supply and use cycle, perform many valuable functions, namely to confront water users with the costs of providing water, to help signal water suppliers when supply augmentation is needed, and to help shape a rational approach to a healthy water environment. Hopefully, equity will be served by appropriate 'life-line' rate structures.

Many pricing improvements need to be made. Wastewater disposal needs to be priced more in keeping with the total costs that pollution imposes on other water users and on riparian ecosystems. City water users should be metered and, in most cases, charged according to an increasing block rate schedule that reflects near term future raw water development and opportunity costs as well as treatment and distribution costs.

The regulations that often constrain water supply agencies in setting their prices need to be reconsidered to allow more appropriate levels of prices and greater freedom to change price under changing conditions. Many utilities have a 'zero profit' constraint that means that, during drought, raising prices is ruled out since inelastic demands imply rising revenues while operating costs may be falling. If the zero profit constraint is too narrowly interpreted, it may also rule out building up sinking funds for system replacements and for future expansions. For many cities in the western US which acquired their raw water supplies (water rights) long ago at low cost, the 'zero profit' constraint means that the current opportunity cost of the raw water being used cannot be reflected in the water prices charged. City accounting practices should be changed to carry those water rights on the books as assets at current value and to count interest charges on the value of those rights as a cost to be recovered through the price structure.

A final needed change relates to overcoming the effects of inappropriate under-pricing of raw water under long-term contracts that cannot be changed for long periods of time. Naturally, increases in these contract prices will be resisted by the beneficiaries of the under-pricing. However, 'water markets' will allow current water contract holders to decide whether or not to sell some or all of their water to higher-valued uses.

There are many such 'win-win' opportunities in water pricing and marketing arrangements today in many countries.

Note

[1.] Toxics, heavy metals, etc. should be prohibited.

References

American Water Works Association (1992) *Water Industry Data Base: Utility Profiles* (Denver: American Water Works Association).

Billings, B. & Agthe, D. E. (1980) Price elasticities for water: a case of increasing block rates, *Land Economics*, February, pp. 73–84.

Boland, J.J., Dziegielewski, B., Baumann, D. & Optiz, E. (1984) Influence of price and rate structures on municipal and industrial water use. A report submitted to the US Army Corps of Engineers, June (Carbondale, IL: Institute of Water Resources).

Crew, M. A., Fernando, C. S. & Kleindorfer, P. R. (1995) The theory of peak-load pricing: a survey, *Journal of Regulatory Economics*, 8, pp. 215–248.

Espey, M., Espey, J. & Shaw, W. D. (1997) Price elasticity of residential demand for water: a meta-analysis, *Water Resources Research*, 33(6), pp. 1369–1374.

Foster, H. S., Jr. & Beattie, B. R. (1979) Urban residential demands for water in the United States, *Land Economics*, 55(1), pp. 43–58.

Hall, D. C. & Hanemann, W. M. (1996) Urban water rate design based on marginal cost, *Advances in the Economics of Environmental Resources*, 1, pp. 95–122.

Howe, C. W. (1982) The impact of price on residential water demand: some new insights, *Water Resources Research*, 18(4), pp. 713–716.

Howe, C. W. (1994) The value of water supply reliability in urban water systems, *Journal of Environmental Economics and Management*, 26, pp. 919–930.

Howe, C.W. & Goemans, C. (forthcoming) the effects of economic and demographic conditions on the functioning of water markets: a comparative study of the benefits and costs of water transfers in the South Platte and Arkansas River Basins in Colorado, submitted to the *Journal of the American Water Resources Association*.

Howe, C. W. & Linaweaver, F. P., Jr (1967) The impact of price on residential water demand and its relation to system design and price structure, *Water Resources Research*, 3(1), pp. 13–32.

Howe, C. W. & Smith, M. G. (1993) Incorporating public preferences in planning urban water supply reliability, *Water Resources Research*, 29(10), pp. 3363–3369.

Inter-American Development Bank (2001) in: C. S. Russell, W. J. Vaughn, C. D. Clark, D. J. Rodriguez & A. Darling (Eds) *Investing in Water Quality: Measuring Benefits, Costs and Risks* (Washington DC: Inter-American Development Bank).

Michelsen, A. M., McGuckin, J. T. & Strumpf, D. (1999) Effectiveness of residential non-price water conservation programs, *Journal of the American Water Resources Association*, 35(3), pp. 593–602.

National Research Council (2002) Committee on Privatization of Water Services in the United States, *Privatization of Water Services in the United States: An Assessment of Issues and Experience* (Washington DC: National Academy Press).

Nieswiadomy, M.L. (1990) Estimating urban residential water demand: the effects of price structure, conservation and education. Paper presented at the Western Economic Association Meetings, San Diego, June.

Renwick, M. & Archibald, S. (1997) Demand side management policies for residential water use: who bears the conservation burden? Paper presented at the annual meeting of the Western Regional Science Association, Big Island, Hawaii, February.

Renwick, M. E. & Green, R. D. (2000) Do residential water demand side management policies measure up? An analysis of eight California Cities, *Journal of Environmental Economics and Management*, 40, pp. 37–55.

Renzetti, S. (1992) Estimating the structure of industrial water demands: the case of Canadian manufacturing, *Land Economics*, November, pp. 396–404.

Schneider, M. L. & Whitlatch, E. E. (1989) User-specific water demand elasticities, *Journal of Water Resources Planning and Management*, July, pp. 789–797.

Scott, A. & Coustalin, G. (1995) The evolution of water rights, *Natural Resources Journal*, 35(4), pp. 821–980.

Tate, D. M. (1990) *Water Demand Management in Canada: A State-of-the-Art Review*, Social Science Series No. 23 (Ottawa: Inland Waters Directorate, Environment Canada).

Tyler, D. (1992) *The Last Water Hole in the West: The Colorado-Big Thompson Project and Northern Colorado Water Conservancy District* (Niwot, CO: University Press of Colorado).

Problems with Private Water Concessions: A Review of Experiences and Analysis of Dynamics

EMANUELE LOBINA

Public Services International Research Unit (PSIRU), University of Greenwich, London, UK

Introduction

Since the early 1990s, private sector participation (PSP) in water supply and sanitation has been introduced into a number of transition and developing countries (Nickson, 1996, p. 2; Braadbaart, 2001, p. 5). The advocates of private sector involvement have argued that PSP would improve efficiency, enable the extension of water services, raise the necessary investment finance, and relieve governments from budget deficits, as summarized by Idelovitch & Ringskog (1995, p. 1). According to the prevailing wisdom, the success of PSP in delivering the expected efficiencies and social benefits depends on the introduction of appropriate incentives and effective transfer of risk to the private sector, as well as adequate reform of governance. While incentives and risk allocation are expected to unleash the efficiency of the private sector, the reform of governance, for example through the creation of an adequate regulatory framework, should safeguard the public interest and ensure a level playing field between the parties involved.

This paper looks at the evidence on the results associated with PSP in water supply and sanitation, which increasingly indicates a failure to deliver the forecast benefits, while eliciting political resistance (Hall *et al.*, forthcoming), and discusses the structural factors preventing PSP from delivering the efficiencies and social benefits expected in theory. Empirical evidence is drawn from transition and developing countries, with particular reference to Latin America. Cases from France are also treated, to introduce and illustrate relevant trends observed at global level, because the French model of delegated management is the prevailing form of PSP in any region of the world, and French-based companies are also dominating the global water industry.

Dynamic Interest-seeking Behaviour

Many commentators take the view that the crucial element in ensuring the success of PSP is the contract itself, which defines rules of subsequent behaviour. Such analysts then explain failure to deliver as deficiencies in the contract, for example in respect of incentives or regulation. Failure to deliver expected results is thus attributed to weakness in contract design, and to be avoided in future by better contractual provisions. As summarized by Braadbaart (2001, p. 6), "[t]he idea was that PPPs would create competition for contracts, that is, competition for the market rather than in the market. Also, by writing elaborate penalty and reward systems into the contract, PPP supporters expected that these would mimic efficiency-driving rivalry in the market place".

By contrast, it is proposed that this discrepancy between practice and theory is best understood by analysing PSP as a dynamic process, with interaction among different actors pursuing different objectives. The most important factor driving outcomes appears to be continual profit-seeking and risk-avoiding behaviour of international water companies, in interaction with local and national governments (pursuing mixed political and fiscal goals), political and community movements, and international donors and institutions pursuing their own goals. The results of this process are strongly affected by the unequal distribution of resources and skills between the parties and by the limited competition in this sector. It is these dynamics, it is argued, which explain the actual (mis)allocation of risk, the (in)effectiveness of governance, and the content (and constant revision) of the contracts themselves, as well as the actual outcomes in terms of investment finance, extension of systems (or failure to extend), pricing policies, and transparency.

For the purposes of this paper, the term 'privatization' is used to mean the partial or total transfer of managerial control of a water undertaking from the public sector to a private operator, usually a TNC (transnational corporation). This definition encompasses a number of arrangements ranging from management contracts to leases/concessions and full divestiture. It broadly coincides with the concept of PSP and better reflects general usage than the definition which restricts it to 'full divestiture only'. It reflects the view that the important element in privatization or PSP is the introduction of a profit-seeking company, whereas the formal variations in the route adopted are less significant in understanding and explaining the actual processes.

Part 1. PSP in Practice: Incentives, Risk Allocation and Governance

PSP and Restricted Competition

A substantial literature relies upon competition to unleash the efficiency of the private sector and ensure the success of PSP in water supply and sanitation, for example through tariff reductions following competitive tendering (Webb & Ehrhardt, 1998).[1] According to Franceys (2000, p. 8) "it is competition that draws the real benefits out of the private sector, just as it draws it from the regulators. Privatization, regulation and competition are the complementary strands of PPP." According to Lorrain, regulation by the markets in the

French system of delegated management necessarily implies competition among urban services groups for entry to local markets (Lorrain, 1997).

Indeed, empirical evidence suggests that competition for entry into the water market is restricted at both global and local level. Outright lack of competition and restricted competition represent an impediment to the delivery of the expected benefits of PSP.

Restricted Competition at Global Level

The global water industry is characterized by a marked concentration, with two TNCs (Veolia and Suez)[2] dominating almost 70% of world private market, joint ventures between these few dominant companies, and difficulty of entry. A similar pattern of concentration, joint ventures and difficulty of entry is also characteristic of the water market in France, which is the home base of the dominant multinationals and of the system of privatization by delegation which has been the core means of privatization in this sector.

Figure 1 compares the 2001 sales of these two and the next largest companies. The process of concentration continued in 2002, notably with a series of takeovers of US companies. In December 2002, Veolia was reportedly considering a bid for Anglian Water's international division which had been put up for sale in October 2002.[3]

Difficulty at entry in the global water industry. The lack of contestability of the global water industry contributes to explain the high level of concentration observed. The global water industry appears to be very resistant to new entrants, even in the case of TNCs well established in sectors other than water and irrespective of their financial resources.

Attempts by Azurix, owned by US-based energy company Enron, to break into the market proved a failure. One of the main reasons for this failure were the poor results obtained when bidding against French-based water TNCs Veolia and Suez-Lyonnaise des Eaux, which could count on superior financial clout and could accept initially lower profit ratios in order to win a tender. As a result, in January 2000 Azurix announced it would change strategy, focusing on smaller projects which would appeal less

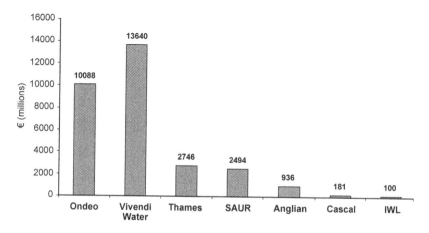

Figure 1. Global private water business compared. *Source:* PSIRU database, derived from companies' annual reports as in Hall (2002a).

to the French giants.[4] In April 2001, Enron announced it had decided to break up Azurix and sell its assets.[5]

In order to establish itself as a global player on the water market, German conglomerate RWE had to take over UK-based Thames Water. RWE had previously established a number of joint ventures with the major water TNCs, but it was only with the takeover of the already established British TNC in September 2000 that RWE acquired a significant share of the world water market.[6]

Restricted Competition at Local Level

Restricted competition also takes the form of joint ventures between TNCs, anti-competitive practices including uncompetitive award of long-term concessions, and endless concessions. All these characteristics are similar to what can be observed in France.

Joint ventures in France and internationally. In France, where they control 85% of private water operations, Suez and Veolia have created joint subsidiaries in a number of towns and regions, with the effect of restricting competition in the French private water market, as ruled by the French competition council in July 2002 (Conseil de la Concurrence, 2002).

This forming of joint ventures is not restricted to France. Figure 2 shows a number of these joint ventures. It is striking that even the nearest competitors to Suez and Veolia, i.e., Thames, SAUR and Anglian, have made partnerships with Suez and Veolia to establish themselves in the market.

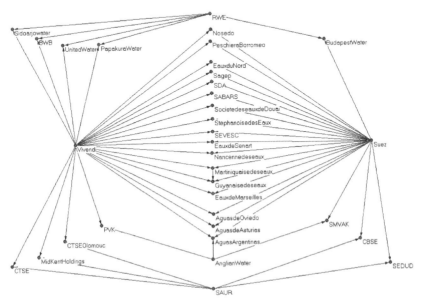

Figure 2. Joint ventures between leading water multinationals, 2002. *Source:* PSIRU database, 2002. Generated by Vladimir Popov, PSIRU using Social Network Analysis software.

Single-bid concessions. Although most contracts are awarded after tendering and competitive bids, there are a number of cases where water concessions have been awarded through one-bidder tenders, or competitive procedures to which only one company put forward a bid.

- *Argentina:* In 1999, an Aguas de Bilbao-led consortium, AGBA, was awarded a 30-year water concession covering one-sixth of the Buenos Aires province. Observers were surprised at the low amount of the bid submitted by the successful consortium, which initially included the Argentine company Sideco. Other interested consortia, including Urbaser and Dycasa failed to submit offers despite being pre-qualified. Accordingly, the Aguas de Bilbao/Impregilo/Sideco consortium submitted the only bid and won the concession in November 1999. Sideco then left AGBA 'for strategic reasons', and was replaced by Dycasa and Urbaser before the concession agreement was finally signed in December 1999.[7]
- *Bolivia:* In September 1999, Aguas del Tunari, a consortium led by International Water Limited (IWL), was awarded a 40-year water supply and sanitation concession for the water and sanitation system of Cochabamba and a complex related infrastructure project called the Misicuni Project. Aguas del Tunari was the only bidder, and its tender was accepted despite serious omissions and irregularities.

Other cases of single-bid concessions in Latin America relate to Thames Water's subsidiary Essam in Chile,[8] Aguas de Barcelona's ACUACAR in Cartagena, Colombia (Nickson, 2001a, pp. 15–16) and International Water in Guayaquil, Ecuador.[9] That pattern is not confined to a specific region, as shown by evidence from Central and Eastern European countries (Hall, 1997; Lobina, 2001).

Endless concessions in France and internationally. The sheer length of water concessions represents a significant obstacle to competition, which can be observed in developed countries with a longer history of PSP in the water sector, such as France and Spain, as well as in developing countries.

In France, until recently, concessions were frequently renewed without tendering, according to the Cour des Comptes, France's national audit body (Cour des Comptes, 1997, pp. 95–97). It was only in 1993 that the so-called Loi Sapin, or anti-corruption law, provided for privatized concessions to be publicly and competitively tendered.[10] In 1995, the so-called Barnier Law limited the maximum duration of privatized water contracts at 20 years.[11] Several contracts were renegotiated and extended a few weeks or months prior to entry into force of the new legislation, sometimes before the contract expiry date (Cour des Comptes, 1997, pp. 105–106). Although in future more concessions can be expected to be re-tendered, in many cases this will not happen before 2020 as contracts concluded before the new legislation became applicable escaped those provisions (Druin, 2002). In Nice, Générale des Eaux has managed water supply and sanitation under a concession contract since 1864.[12] Even when contracts are re-tendered, due to its superior knowledge of the system, the incumbent remains at a competitive advantage in respect of competing companies and this strongly undermines the effectiveness of competition (Cour des Comptes, 1997, pp. 98–99).

In 1902, the city of Valencia, Spain awarded a 99-year water concession to a private company, AVSA. In the late 1990s, the city of Valencia began to draw up tender documents but AVSA, now part of the SAUR-Bouygues group, announced that if it lost the tender it would demand compensation of €54 million Euros for investments it had made in the system.[13] The tender proceeded, with a clause stating that the winner would have to pay €54 million to AVSA. Not surprisingly, there was not a single competing bid. AVSA, now part of a joint venture with the council itself, will enjoy the concession for a further 50 years.[14]

Similar restrictions to competition apply to developing countries and are a cause for concern in the light of the limited experience available with managing long-term water concessions throughout the whole lifecycle of the contract. Few water concessions in developing countries have in fact lasted long enough yet to reach their appointed end. The most long-standing example, in the Côte d'Ivoire, was originally awarded without any public competitive tendering, and was later renewed and extended without any public or competitive tendering (Bayliss, 2001). In 1998, a SAUR-led consortium was awarded a 95-year water concession in Mendoza, Argentina.[15] In June 1999, a Suez-Lyonnaise des Eaux/Agbar consortium bought 42% of Santiago de Chile's water company EMOS. The private consortium was also awarded an unlimited duration concession to manage and develop the city's water and sewerage (Suez Lyonnaise des Eaux, 1999, p. 25).

PSP and Risk Allocation

In theory, PPPs are expected to unleash the efficiencies of the private sector and deliver social and environmental benefits subject to the effective allocation of operating and political risks to the parties best placed to minimize and manage such risks (ADB, 2000, p. xvii). It is generally assumed that private operators are best at dealing with operating risks while public bodies should preferably retain the political risks involved with PPPs.

The World Bank has stressed that whether contractual options ranging from management contracts to full concessions "perform better than full provision by state-owned enterprises depends in particular on whether performance risk is effectively shifted from taxpayers to the private shareholders of the company that enters into a concession-type arrangement" (World Bank, 2002, pp. 23–24).

Performance or operating risk may be identified as the variety of risks associated with the commercial provision of water supply and sanitation services, such as failing to meet the agreed service targets or facing declining revenues as a result of decreasing consumption. Political risk may be defined as the risk faced by a commercial water operator due to undue governmental interference, such as in the case of breach of contract, expropriation, restrictions to currency transfer, or war and civil disturbance.

Performance Risk

Guarantees. Irrespective of the availability of technical instruments designed to shift performance risk on to the private operator, such as performance bonds, water PPPs often appear as virtually risk-free for international operators. Contractual agreements may directly or indirectly guarantee a steady rate of return, or may be renegotiated in case of failure to achieve the originally set investment objectives or service targets.

- *Bolivia:* The Cochabamba water supply and sanitation concession allowed the operating company Aguas del Tunari, a subsidiary to International Water Limited (IWL), to enjoy a guaranteed 15% real return for 40 years (Lobina, 2000).

- *Chile:* In June 1999, a Suez-Lyonnaise des Eaux/Aguas de Barcelona consortium acquired a 42% stake in Santiago water company EMOS. The consortium, which was awarded an unlimited duration concession to manage and develop water and sewerage, reportedly enjoyed a "constant level of profitability, of roughly one-third of total sales, ... guaranteed by the state".[16]

Concession arrangements often guarantee a steady rate of return by shielding the operator from a downfall in revenues due to variations in water consumption. This may take place in the form of clauses providing for automatic tariff adjustment in case of variations in demand or in the form of bulk water supply schemes, requiring the conceding authorities to buy fix volumes of water irrespective of future demand. Guarantees against demand risk thus shift the burden of operating risk on to taxpayers or consumers, depending on whether the guarantees require compensatory payment by local government or tariff adjustments.

Examples of bulk water supply schemes providing for guarantees against demand risk in the form of take-or-pay clauses include Chengdu, China,[17] Thu Duc, Vietnam[18] and Yuvacik, Turkey. The Yuvacik scheme was designed "to provide water for the town of Izmit, as well as an urgently needed extra source of water supply for Istanbul... However, certain industrial users and neighbouring municipalities refused to buy water from the plant as it was too expensive" (Hall & Lobina, 2004, p. 271).

Furthermore, less than accurate or over-optimistic demand projections may trigger renegotiation and alter the originally agreed contractual balance. There are a number of cases where prices have been forced up within a year or two of a concession being started, where the assumption in the original contract about the level of demand turned out to be too high, and the real level of demand was then used as a justification for large price rises.

- *East Germany:* In Rostock, water prices rose by 24% in 1996, 22 months into a 25-year contract, because over-estimated consumption would have led to losses for Suez subsidiary Eurawasser, and so "The shortfall automatically activated price-adjustment clauses within the Eurawasser contract" (Hall, 1998, p. 129).[19]
- *South Africa:* In June 2001, SAUR's subsidiary, Siza Water, renegotiated its contract with the municipality of KwaDukuza in Dolphin Coast, only 2 years and 4 months after the award of the 30-year concession. Siza asked for relief under the contract, as a result of actual demand failing projections. As the expected development of middle-income and mass housing had not materialized, Siza's revenues were affected by a shortfall of some R12 million a year. The renegotiated agreement provided for an immediate 15% price increase and a R15 million reduction in projected investments, from R25 million to R10 million in 5 years.[20]

Faulty demand projections might also affect the sustainability of bulk water supply schemes. In Chengdu, due to over-estimation of demand the take-or-pay clause is forcing public water company CMWSC to cut its water production by 40% and buy 'higher priced water' from a Veolia and Marubeni consortium (ADB, 2004).

Renegotiation and reduction of investments. Modest investment and service targets are instrumental in removing operating risk. Renegotiation in the light of changed operating conditions may either cancel original investment commitments or substantially postpone their realization. There are a number of cases where, shortly after the award of a concession, private operators have revised downwards projected investments.

- *Belize:* In March 2001, the Belizean government privatized the Belize Water and Sewerage Authority (WASA) through the sale of shares.[21] Within weeks of the deal, the private operator announced that sewerage connection charges would increase from $80 to over $1000 and that "it would not spend the $140m which it had promised on new capital investment because the company had been 'tricked' in the fine print of the original purchase agreement".[22]

- *Aguas Argentinas:* In May 1993, a Suez-Lyonnaise des Eaux-led consortium started operating a 30-year water supply and sanitation concession in Buenos Aires, Argentina. It was only eight months later that the operating company Aguas Argentinas requested an "extraordinary review" of tariffs, due to unexpected operational losses. The concession agreement was then renegotiated from February to September 1997 (Azpiazu & Forcinito, 2002, pp. 23–28). A recent study has estimated that from May 1993 to December 1998, Aguas Argentinas failed to realize 57.9% of the originally agreed investments for a total of US$746.39 million (see Table 1) (Azpiazu & Forcinito, 2002, pp. 44–45).

Similar cases include Cordoba (Nickson, 2001b, pp. 1, 14–15) and Santa Fe (see Table 2 below), Argentina and Campo Grande,[23] Brazil.

Weakness of performance bonds. Performance bonds, deposited by private concessionaires to guarantee achievement of agreed performance targets, are the normal way of providing protection against non-performance. However, governments may prove reluctant to call in these bonds for a number of reasons, including unwillingness to engage a dispute with a powerful TNC—which may be withholding payment of lease or concession fees to add pressure, concerns for the continuity of service provision, and the costs of re-tendering or re-establishing a public service. This calls into question the effectiveness of performance bonds and general ability of local governments to force private operators to assume risk for poor performance.

- *Philippines:* Water supply and sanitation in Manila were privatized in January 1997 to two private groupings: a Lyonnaise des Eaux-led consortium to operate in the western zone of the city and a North West Water-led consortium to operate the eastern zone (Esguerra, 2002, p. 1). Suez subsidiary Maynilad failed to tackle Unaccounted-For-Water (UFW) in west-Manila. Despite pledging to cut UFW from 57.4% to 42.0% between 1997 and 1999, Maynilad saw UFW rise from 63.3% to 67%. IWL subsidiary Manila Water Company, operating in east-Manila, projected a reduction in UFW from 44% to 22% between 1997 and 1999. In fact, UFW decreased from 45.2% to 39.8% (Esguerra, 2002, p. 31).[24] Each of the two concessionaires deposited a US$200 million performance bond, which the government failed to recall. By contrast, in September 2001 the government agreed to renegotiate the terms of the two concession agreements and allow the companies recover the losses suffered for having submitted unrealistic bids and caused by currency devaluation (Esguerra, 2002, pp. 5, 17, 33).

- *South Africa:* In February 1999, SAUR subsidiary Siza Water was awarded a 30-year contract to provide water supply and sanitation to the resort of Dolphin Coast, South Africa. In June 2001, Siza Water requested renegotiation of the contract after being hit by financial problems. Faced with a serious shortfall in

Table 1. Investment and under-performance by Aguas Argentinas SA, 1993–1998 (in millions of pesos/dollars at supply values)

	1993	1994	1995	1996	1997	1998*	Total
Committed to investments in original bid	101.5	210.52	302.91	362.36	229.10	83.07	1289.46
Investments realized	40.93	144.55	132.17	100.49	109.52	15.41	543.07
Under-performance	−60.57	−65.97	−170.74	−261.87	−119.58	−67.66	−746.39

Note: *Corresponding period: May–December 1998.
Source: Azpiazu & Forcinito (2002, p. 45); elaboration on the basis of data provided by the Users' Committee of ETOSS.

Table 2. APSF (Santa Fe, Argentina), amount provided for investments in original concession agreement and in second renegotiation (proposed renegotiation, not implemented) (in millions of pesos/dollars)

Five-year period	Amount provided for investments in original contract	Amount provided for investments in second renegotiation
1996–2000	290.00	245.00
2001–2004	211.00	80.00
2005–2008	206.00	80.00
Total	707.00	405.00

Source: Muñoz (2002).

revenues of about R12 million a year, due to demand failing to reach projected levels, Siza Water refused to pay the scheduled R3.6 million lease payment due to the municipality of KwaDukuza and obtained the renegotiation of the contract in its favour. According to KwaDukuza's acting municipal manager, the only alternative to renegotiation was "to go off to the contract guarantor (a bank) and take back the performance bond".[25]

Allocation of Currency Risk in Practice

In order to attract investment finance, PPPs have to meet the requirements of financial investors who have their reluctance to accept risk, especially currency risk. Water multinationals naturally seek to reduce currency risk, and do so by seeking provisions for indexing tariffs to the value of US dollar, in order to preserve the real value of profits generated by local operations and of payments in hard currency to subcontracted subsidiaries.

Whatever the provisions governing private water operations, the reality appears to be a contest between the parties: in Manila the original provisions implied recovery of currency losses in the future (Esguerra, 2002, pp. 11–12), but the operators were not prepared to accept this delay and sought to insist on earlier recovery (Esguerra, 2002, p. 20).

In Buenos Aires, where renegotiation introduced clauses apparently providing for 100% protection against currency risk, through the 'dollarization' of tariff indexing, the collapse of the Argentine economy following the financial and political crisis of December 2001 effectively nullified these guarantees which became unsupportable so that the new government sought their cancellation (Azpiazu & Forcinito, 2002, pp. 53–65; Hall and Lobina, 2004). In effect, currency risk was not avoided, simply converted into political risk as Aguas Argentinas would resort to international arbitration against the Republic of Argentina.[26]

Political Risk and Implications of Distorted Risk Allocation

Facing limited operating risk, private water operators have little incentive to deliver the expected efficiencies and performance achievements. Facing little currency risk, water multinationals have no incentive to achieve efficiencies through reducing the purchase of US dollar-denominated goods and services bought from their own subsidiaries, in favour of purchasing local inputs.

Distorted risk allocation not only alters incentives for the private operators, it may mean that PPPs turn into debt-like obligations for public authorities and, more broadly, local communities. The public's financial obligations to pay charges for the private operation for 20 years or more remains constant, while the operator has an incentive to and often does negotiate away the original commitments in terms of service, investment and/or price levels. Resort to commercial and legal instruments to protect operators' interests against political risks may then place additional costs on to local taxpayers. From a private operator's point of view, this risk allocation is a satisfactory outcome for profit maximization, but at the cost of undermining the social and political sustainability of PPPs. PPPs which so favour private concessionaires might end up triggering political risk rather than taking advantage of risk mitigation and produce the expected developmental results.

This can be seen in a number of cases where poor performance or excessive prices have sparked political decisions to terminate privatized concessions and operating contracts. In most cases, those events were prompted by consumers' resentment against what was perceived as an unsustainable situation. In turn, water TNCs have filed multi-million compensation claims in front of international arbitration panels. The exit from privately-run operations thus proves either difficult or costly for the local political authority and community, a prospect which decision makers need to bear in mind when considering the potentially adverse long-term implications for host countries and local communities resulting of entering PPPs in the water sector.

- *Tucuman, Argentina:* In May 1995, Aguas del Aconquija, a subsidiary to Générale des Eaux, was granted a 30-year concession to supply the province of Tucuman. Although water tariffs doubled following the award, the company reportedly "made good on few promised investments during its tenure".[27] When water became 'inexplicably' brown, more than eight out of 10 consumers stopped paying their bills and the concession was terminated in October 1998. The company filed a US$300 million compensation suit with the World Bank's ICSID (International Centre for Settlement of Investment Disputes), and in November 2000 the international arbitration court decided to dismiss Veolia's claims (ICSID, 2000). The French multinational has appealed (Gaillard, 2003).[28]

Similar cases include Cochabamba, Bolivia (Lobina, 2000), [29] Buenos Aires Province, Argentina (Hall & Lobina, 2002, pp. 13–14) [30], Potsdam, Germany[31] and Nkokobde, South Africa[32]. In addition, the case of Szeged, Hungary shows how international arbitration and any other legal confrontation potentially implying payment of multi-million compensation might affect local authorities' decisions on exit from unsatisfactory contracts.

- *Szeged, Hungary:* International arbitration was a deterrent against the termination by Szeged city council of a flawed operating concession to Veolia. The dispute arose from the determination of the local government not to allow further increases in operation and maintenance costs. Exclusive subcontracting to a Veolia subsidiary was a major determinant of inflated costs and allowed the French TNC to export a high share of the profits (Lobina & Hall, 2000). As Générale des Eaux repeatedly refused to renegotiate the concession agreement, the municipality of Szeged established its own water company in July 1999 and decided to terminate the contract with Générale des Eaux. The dispute was amicably settled in February 2001 after Vivendi Water had resorted to international arbitration.[33]

PSP and Governance

The theory of PPPs in the water sector does not only rely on competition and risk allocation for the private sector to unleash its efficiency and deliver the expected social and environmental benefits. It also relies on the reform of governance, for example through the introduction of regulation of prices and outputs or the monitoring by local government retaining a shareholding in the operating company. The rationale for regulation in water supply and sanitation is provided by the monopoly structure of the industry (Franceys, 2000), implying the risk that the private operator might otherwise abuse its monopoly power. Regulation is thus viewed as an imperfect substitute for competition aiming at safeguarding consumers' interests while providing private companies with the incentives to invest and operate efficiently (Klein, 1996; Rees, 1998).

One of the constraints to regulation identified by the literature is the asymmetry of information between the regulator and the regulated operator, that is to say the fact that regulators "tend to be less well informed about costs and quality of water systems operations than the water company management" (Klein, 1996, p. 9). The following empirical evidence confirms the view that asymmetry of information reduces the effectiveness of regulation. There is evidence that asymmetry of capacity is also a problem, as regulators may simply lack the resources of the private companies in addressing any areas of conflicting interests. Convincingly, regulation is described as a bargaining process between the parties involved, whose outcome is expected to "depend on the resources (power) and needs of the various players", who include a broad range of stakeholders (Rees, 1998, p. 27).

Access to and Asymmetry of Information

The effectiveness of regulation and monitoring mechanisms is affected by the ability of local governments and communities to access vital information on the operations carried out by private operators, as well as by the asymmetry of information between regulators and regulated operators.

• *Commercial confidentiality:* Commercial operations invariably prefer confidentiality and secrecy, as it protects their ability to manage financial affairs to maximize the benefit to their owners. Aguas del Tunari, the privatized water concession in Cochabamba, Bolivia, refused to disclose the financial model behind its price rises on the grounds that the model itself was a commercial secret.[34] In Fort Beaufort, South Africa, the contract prevented any member of the public from seeing the contract without the explicit approval of the company WSSA (owned by Suez-Lyonnaise).[35] Documents relating to the privatized Budapest Sewerage Company, operated by Veolia, are kept secret, even from council officials, and Budapest City Council debates related issues only in closed sessions[36] (Hall, 2002a).

• *The cost of asymmetry of information:* In 1989, a consortium led by SAUR and Veolia was awarded a lease contract to operate and manage urban water services in 17 urban centres in Guinea. The private consortium set up SEEG as the operating company. A report had pointed to the fact that state water authority SONEG did not have access to information on SEEG's finances. This meant that when SEEG put in a request for SONEG to increase tariffs, SONEG was not able to accurately assess the grounds on which this was based. A World Bank audit found that weak regulation meant that the formulas used to adjust prices in response to cost changes were misapplied so that tariffs were overvalued (Bayliss, 2001; Ménard & Clarke, 2000, p. 45).

Capacity to Regulate

The asymmetry of capacity can be observed not only in developing countries but in France itself, the home of the PSP model and the dominant TNCs. In 1997, a report by French audit body Cour des Comptes (1997) repeatedly emphasized the disparity between the local authorities and the three giant companies dominating the national water market, which resulted in inadequate monitoring of privatized operations. *Le Monde* commented that the French system of delegated management "left elected councillors on their own, without support, to deal with conglomerates wielding immense political, economic and financial power".[37]

- *Newly Independent States (NIS):* In October 2000, the NIS Ministerial Conference on Water Management and Investment acknowledged that: "The capacity of most NIS governments to effectively regulate private sector participation, particularly the more extensive forms, is an important constraining factor [on developing public-private partnerships]". [38]
- *South Africa:* A recent study of public-private-partnerships in South Africa concluded that "lack of public sector capacity, which is often offered as a reason to outsource to the private sector was found to be, in this case, an important reason *not* to privatize" (Bakker & Hemson, 2000, p. 9).
- *Colombia:* A recent study on the PPP set up in Cartagena by Suez concluded that: "The Municipality of Cartagena lacks the minimal technical support in its negotiations within the joint venture. To all intents and purpose it is a 'sleeping' partner" (Nickson, 2001a, p. 34).

Lack of Transparency and Accountability

Lack of transparency and poor institutional development are key factors in determining the failure of regulation (Nickson, 1996).

- *Argentina:* The Argentine government has bypassed local water regulators during the renegotiation of the concessions held by Suez-Lyonnaise des Eaux subsidiaries. A Ministerial Decree provided for the Secretary of Natural Resources and Sustainable Development to replace Buenos Aires regulator ETOSS in the renegotiation of the concession with Aguas Argentinas "given the technical complexity of the issues relating to the regulation" of the service (Azpiazu & Forcinito, 2002, p. 26). Similarly, during the renegotiations with Aguas Provinciales de Santa Fe, regulator ENRESS was sidelined by the Sub-Secretary to Public Services (Muñoz, 2002).

The case of Manila provides further examples of how a regulatory agency might be subject to external pressures due to the legal (i.e. contractual) nature of the establishing act (Esguerra, 2002, pp. 17–18) or the (poor) level of job security enjoyed by the individual members of the regulatory body.[39] The independence of regulatory institutions may be further affected by their lack of autonomous financial resources. In many cases, in order to alleviate governmental budgets, regulatory agencies rely on a fix percentage of tariffs collected by private water operators. As a result, regulators might appear as permeable to the interests of water multinationals.

- *Renegotiation and regulation as bargaining in Manila:* In September 2001, Maynilad's concession agreement was renegotiated providing for the Suez subsidiary to "resume payment of maturing concession fees, at least MWSS' current operating budget", in exchange for tariff adjustment to currency devaluation. The renegotiated agreement also established that "Maynilad shall withdraw its case filed against MWSS and in turn, MWSS shall suspend calling on the performance bond posted by Maynilad" and paved the way for the postponement of water pressure obligations and of investments in sewerage (Esguerra, 2002, p. 33). In November 2002, regulator MWSS-RO was in the process of resetting tariffs after the first five years of operations as provided for by the concession agreements. After defining the new tariffs, the regulator would set performance targets in consultation with the two concessionaires "since these targets should be mutually agreed upon".[40]

Corruption, State-capture and Governance Failure

For a private water company, the incentive to engage in corruption of public officials is not only to bypass competition but also to seek more favourable contract terms, and laxer regulation or monitoring. The local community thus has to bear the costs not only of the bribe but also of the distorted terms of contract and regulation.

- *Grenoble:* French courts of justice found that the Grenoble water service had been privatized in exchange for contributions by Lyonnaise des Eaux to the mayor's electoral campaign, and other gifts, totalling over FF19 million.[41] In 1995, the regional auditor found that over the lifecycle of the contract the costs borne in excess by local consumers and taxpayers would be more than FF1 billion (US$150 million) (Hall & Lobina, 2001).[42]

Allegations of corruption have been made in many cases of water privatization in developing countries, but without criminal convictions of water TNCs' executives. This is consistent with research by a division of the World Bank itself, which "found that transnational firms are just as likely to pay administrative bribes and to try to capture the state as other firms, and that transnational firms headquartered abroad are more likely than other firms to pay public procurement kickbacks" (Hellman *et al.*, 2000, p. 5).

Cases of corruption associated with PSP in the water sector, in developed as well as developing countries, are treated in Hall & Lobina (2004, pp. 267–268) and Hall (1999).

Weak Accountability and Lack of Public Participation

In a context of less than transparent decision making on the award of privatized water concessions and weak accountability, public participation would have a role to play in institutional development. In fact, despite the emphasis placed on community participation by the advocates of PSP, effective public participation in decision making on and monitoring of PSP is either non-existent or ineffective. A review of community participation in water projects found that: "It is uncommon to find a project with a large private and community involvement component" (Miller, 1999).

The following example shows how PSP in the water sector is associated with weakened accountability and restricted opportunities for public participation in decision making.

- *Nkokobde, South Africa:* The court who gave the ruling on the cancellation of the contract between the municipality of Nkokobde and WSSA found the contract was invalid as it had not been published first for comment by members of the public. Second, approval from the local government MEC was never obtained.[43]

Further examples include the secrecy surrounding the documents relating to the privatized Budapest Sewerage Company and restricted debates on its operations (see above, 'Access to and Asymmetry of Information') and the decision to award the Casablanca, Morocco water and energy concession to Suez (Hall, 2001).

Part 2. PSP in Practice: Costs and Outputs

PSP and the Costs of Investment Finance

In a context of growing urbanization, large-sized water operators, and TNCs in particular, are expected in theory to reduce the costs of investment finance. Lorrain (1997) elaborates on the comparative advantages enjoyed by large water operators on the basis of economic and organizational arguments. The economic arguments relate to economies of scale and the corresponding increase in yield rates, and to economies of scope. The industrial organization arguments relate to transaction costs and optimal technical systems. As regards transaction costs, "The organization of a large group that relies on a great number of subsidiaries and on networks of subcontractors can, by performing work in-house, avoid using the market, with its scheduling constraints, myriad of controls and associated extra costs. These internally-driven organizations, contractually flexible, constitute for some activities a viable alternative to the market outcomes" (Lorrain, 1997, pp. 24–26).

Empirical evidence suggests that these comparative advantages enjoyed by water TNCs in terms of scale, scope, transaction costs, and other size- and capacity-related aspects do not translate in lower costs for local consumers and taxpayers.

PSP, Risk and Transaction Costs

Transaction costs can be defined as the legal, consulting and financial costs of structuring an infrastructure project. To these costs, typically incurred by private project developers, it is also necessary to add transaction costs borne by host governments, for example the costs of setting up and running a regulatory agency. Despite variations in transaction cost levels across countries and different sectors, a World Bank study has identified transactions costs in private infrastructure projects as possibly ranging between "some 5 to 10 percent of total project cost" (Klein *et al.*, 1996, p. 1).

In relation to PSP in water supply and sanitation, transaction costs can be seen as the costs of identifying, allocating and mitigating the various risks involved. Transaction costs are clearly higher for PSP than for water operations under public ownership. One reason for this is that there is no need for such complex legal provisions, as the public authority is unavoidably responsible for ensuring water supply, whatever happens, and so the liabilities of each side do not need to be spelt out in advance. Some transaction costs only exist under PSP: for example, political risk, defined as "the probability of the occurrence of some political event that will change the prospects for the profitability of a given investment" (West, 1996, pp. 6–7), only exists under PSP and therefore does not apply to public water operations.

Advocates of PSP argue that the higher transaction costs under private sector projects are due to the fact that risks perceived by private developers remain hidden under public provision and ultimately borne by consumers or taxpayers (Klein *et al.*, 1996). For example, political risk might translate into political interference or inefficiency under public water operations. However, this argument relies on an unsupported assumption that political decision making in this area is always associated with dysfunctional results, which leads to a consistent relative inefficiency of public sector operators compared with private sector operators. However, the empirical evidence does not support this assumption about inefficiency: the results of most empirical studies, in water as in other sectors, are, overall, neutral on whether public or private ownership is more efficient (Lobina & Hall, 2000; Hall, 1998, 2002b). The empirical evidence supports this position in all sectors, not only water, as well as a general case that political decision-making actually increases efficiency in monopoly sectors (Willner, 2001).

There is a further layer of transaction costs involved in PSP, which is the restructuring of the water sector to make it suitable for privatization through a concession model, often through legislation. These and other transaction costs are often substantial enough to require specific development bank loans to cover these costs alone.

- *Paraguay:* In July 2000, the World Bank was to issue a US$12 million loan to Paraguay aiming to finance the preparation costs of the privatization of state water company Corposana and telecoms company Antelco. Would-be private operators were expected to refinance the loan.[44] In June 2001, the Paraguay government and the World Bank awarded a US$2 million contract to Spanish firm Sanchis & Asociados to provide an 18-month public relations campaign supporting the privatization of Corposana and Antelco. This was the largest PR contract ever awarded in Latin America.[45]
- *Guayaquil, Ecuador:* In July 1997, the IBD approved a US$40 million loan aiming to improve water supply and sanitation services in Guayaquil. In preparation to privatization of water operations, US$20.2 million out of the US$40 million lent were earmarked to "undertake technical, legal and financial studies and to prepare bid specifications for the concession award".[46]

PSP, Risk, Incentives and the Cost of Finance

A number of structural factors contribute to inflating the costs of investment finance under PSP, including the profit motive as well as the typical reluctance of private shareholders to assume the financial risks related to a long-term project and secure loan repayment. In the water sector, project finance is the preferred method of structuring investment finance as loan repayment is secured entirely on the cash flow generated by the operation of the project. This means that, contrary to corporate finance where loan repayment is secured on the balance sheet of the operating company's shareholders, under project finance most if not all of the financial risk is shifted on to local consumers or taxpayers.

- *Manila:* Both concessionaires were subject to the pressures resulting of currency devaluation, but Maynilad experienced considerable difficulty in achieving financial closure of a US$350 million loan to finance investments in the first five years of operations. While Manila Water Company resorted to corporate finance, Maynilad opted for project finance or limited recourse financing. Difficulties at reaching financial closure were due to the fact that "Other things being equal, the risk premium that creditors would assign to Maynilad would be higher and this raises the financing cost. Maynilad's creditors would also be far more meticulous than Manila Water's creditors when it comes to rights of third parties (e.g. creditors) to Maynilad's assets and future income streams, especially in case of bankruptcy or default" (Esguerra, 2002, p. 10). It should also be noted that in October 2001, Maynilad's concession agreement was renegotiated and the regulator agreed to address the concerns of lenders, including the postponement of sewerage investments beginning in year five and the postponement of water pressure obligations (Esguerra, 2002, p. 33). The lenders were concerned that the creditworthiness of the Suez subsidiary could be affected by the company's failure to meet its contractual obligations and the eventuality of forfeiting its performance bond (Esguerra, 2002, pp. 9–10).

Overall, the profit motive has a major impact on the long-term costs of PSP, especially in view of pressure from parent companies with double-digit growth projections as is the case with Veolia and Suez.

- *Double-digit growth projections:* In October 2001, Suez CEO Gerard Mestrallet promised shareholders double-digit growth in the water sector for fiscal year 2001–2002 by more than 10%.[47] In May 2002, Mestrallet "confirmed his target of double-digit compound annual average growth in sales and earnings per share in its three global businesses (water, energy and waste) on the 2001–2004 period".[48] In August 2001, it was reported that Vivendi Environnement, active in the water sector as well as in energy, waste and transport, had announced targets of "10% a year growth in turnover, and 14% a year in operating profit".[49]

Internal Subcontracting

Vertically integrated water TNCs have a vested interest in purchasing equipment from and subcontracting services to their own subsidiaries in order to maximize profit from water and wastewater operations to the benefit of shareholders. A private operator's ability to pass the entire costs of subcontracting on to local consumers or, as a last resort, taxpayers, may give rise to such a vested interest. Similar practices affect competition in the acquisition of external services and goods, undermine the operator's incentive to achieve efficiencies and may have a significant impact on operating and capital costs. The implications of privileged or exclusive access to subcontracting can be observed in developed (Hall & Lobina, 2001), as well as in transition and developing countries.

- *Subcontracting and costs in Manila:* The bid successfully submitted by Suez subsidiary Maynilad estimated aggregate operating expenses for 1997, 1998 and 1999 to be PHP4.369 billion. In fact, actual operating expenses for the corresponding period amounted to PHP6.259 billion, that is 43% more than projected. Internal trading might explain at least part of such extra costs. As has been reported in a comprehensive study on the Manila water privatization, (Esguerra, 2002, p. 10) "there are also claims from disgruntled employees at Maynilad and from an MWSS source that the company had very expensive contracts with the affiliate companies of the French and Filipino corporate sponsors". These contracts included a management consultancy contract, interest bearing advances for bidding and start-up costs, technical assistance and service agreements, guarantee fees related to loans and standby letters of credit guaranteed by the two private shareholders, an interim programme management deal, and a technical assistance agreement for reducing non-revenue water. "All of these deals are denominated in foreign currency and thus became inflated as a result of the peso devaluation" (Esguerra, 2002, p. 10).

Other cases of subcontracting to subsidiaries of the same TNC involved in local water supply and sanitation operations include Szeged, Hungary (Lobina & Hall, 2000) and Łodz, Poland (PSPRU, 1998). There is also a striking coincidence between the award of a number of privatized water concessions and the subsequent subcontracting of major works contracts to subsidiaries of the same water TNCs, as exemplified by the Chilean experience of Anglian Water[50] and Suez.[51]

However, it should be noted that water TNCs might succeed in extracting increasing resources out of a given concession through techniques other than subcontracting to their own subsidiaries. Such techniques might include the award of management contracts to subsidiaries of the same mother company, providing for payment of ever increasing management fees, as was the case with Aguas de Barcelona's experience in Cartagena, Colombia (Nickson, 2001a).

PSP and the Dynamics of Water Pricing

Post-award Dynamics

The advocates of PSP in water supply and sanitation argue that competition for the market, for example in the form of competitive bidding for long-term concessions, delivers benefits to consumers as it would result in 'spectacular' tariff reductions in respect of previous public water operations. Examples referred to in support of this view include the experience in countries such as Argentina and Guinea as well as in Manila, Philippines (Webb & Ehrhardt, 1998).

Empirical evidence seems to contradict the above static interpretation of water pricing under PSP as, particularly as a result of private operators' interest-seeking, pricing proves to be a dynamic process. Once private operators are isolated from competition, that is to say immediately after the beginning of operations, the same factors observed as affecting the functioning of PPPs will also determine pricing and indeed the broader outcomes of PSP. In this sense, restricted competition, distorted risk allocation, contract renegotiation and regulation in a context of unequal distribution of resources and capacity, not to mention lack of transparency, often mean that the initial tariff reductions are only temporary and destined to be more than offset by successive increases.

Another aspect of pricing as a dynamic process is the relative impact of tariff increases, possibly exceeding the extent of price variations in nominal terms, which might result from concurrent reductions in the level of projected investments or in other outputs. Before presenting empirical evidence on the dynamic factors conditioning pricing, it should also be noted that constraints affecting policy and decision making on urban water systems reform have repercussions on the dynamics of water pricing. For example, tariff reductions following the award of privatized concessions may seem less spectacular when considered that, due to ideological or other political motivations, rates charged by public undertakings may be increased substantially to make forthcoming privatizations more appealing to potential investors. In the light of the complexity of privatization as a political process, the dynamics of pricing under PSP might in fact stretch to the preparatory period preceding the competitive bidding.

Making Privatization more Appealing

It is often the case that local governments decide to increase water tariffs charged by public providers as a way of attracting greater interest from international operators and investors and facilitate the public perception of the privatization as a political success. Successful bidders are thus in a position to substantially reduce tariffs, which may then start to increase again immediately after the beginning of privatized water operations. Tariff reductions as a result of international competitive bidding may also appear less impressive, if not illusory, when considering the favourable contractual conditions obtained by the private operator, in terms of either the expansion of the volume of water and wastewater on which the tariff is applied, or in terms of commitment to achieve quality standards as it appears to be the case of the Province of Santa Fe, Argentina (Muñoz, 2002).

- *Buenos Aires, Argentina:* Water tariffs in Buenos Aires were raised a number of times prior to the award of the 30-year concession to Aguas Argentinas in May 1993. In February 1991, tariffs were augmented by 25%, followed by a 29% increase in April 1991. In April 1992, tariffs were marked up by an 18% VAT charge and a further 8% tariff increase was introduced just before May 1993. The Suez-Lyonnaise des Eaux-led consortium won the concession as it offered the highest reduction in tariffs, that is to say a 26.9% downwards adjustment (Azpiazu & Forcinito, 2002, pp. 7, 12). However, the concessionaire immediately started to renegotiate the agreement in order to obtain price increases and the deferral of its investment plan. From May 1993 to January 2002, bills increased by 88.2% (see below for further details 'Renegotiation and Dynamic Pricing') (Azpiazu & Forcinito, 2002, pp. 38–39).
- *Manila, Philippines:* In August 1996, tariffs charged in Manila by publicly owned MWSS were increased by 38%, from PHP6.43 pcm to PHP8.78 pcm. Following the January 1997 privatization, tariffs charged by Maynilad in west Manila fell by

43.5%, down to PHP4.96 pcm, while tariffs charged by Manila Water Company in east Manila fell by 74%, down to PHP2.32 pcm (Esguerra, 2002, p. 4). It is not clear to what extent successive tariff increases agreed upon to compensate for the losses suffered by the two concessionaires reflected the scale of currency devaluation. As of March 2002, prices charged by Maynilad reached PHP15.46 pcm, while prices charged by Manila Water Company reached PHP6.75 pcm (Esguerra, 2002, p. 34). Esguerra (2002, pp. 2, 24, 32) argues that currency devaluation might only explain part of the losses incurred by the two concessionaires, which were also due to inefficiency and over-optimistic assumptions.

Currency Risk Allocation and Pricing

The most obvious examples of risk allocation affecting pricing relate to currency risk. Profit motive and the typical reluctance of private investors to assume currency risk also affects the costs of investment finance ultimately borne by local communities. Water TNCs appear to have no vested interest in seeking the removal of currency risk away from local consumers and taxpayers and often require the indexation of tariffs to the US dollar and/or US inflation. In fact, profits from water operations are repatriated in US dollars or other hard currency and failure to provide for tariff indexation would deteriorate commercial returns. Subsidiaries of water TNCs also tend to purchase goods and services from other subsidiaries of their own group, often at a high cost, and those goods and services are paid for in US dollars.

- *Buenos Aires, Argentina:* The Aguas Argentinas concession agreement was renegotiated for the second time from February to September 1997 and, among other provisions, the new renegotiated contract provided for automatic tariff adjustment in case of currency devaluation. In July 1999, the regulatory and pricing framework was modified to adjust tariffs to the price index IPD. However, instead of adopting a domestic price index, IPD was an average between the US Producer Price Index and Consumer Price Index, both of which would then increase more than domestic prices (Azpiazu & Forcinito, 2002, pp. 25–31). A further attempt to link local water tariffs to variations in US prices was made in January 2001, when Aguas Argentinas and ETOSS agreed on an aggregate 3.9% tariff increase for the years 2001 to 2003 "to finance the company's US$1.1 billion 1999–2003 investment plan", together with a 1.5% adjustment to US inflation. The tariff adjustment to US inflation was introduced although the Argentine economy was experiencing deflation. The agreement also provided that starting from 2003 water rates would be indexed to US inflation.[52]
- *Cochabamba and La Paz, Bolivia:* A recent study revealed that both the 1997 concession awarded to Aguas de Illimani in La Paz and the 1999 concession awarded to Aguas del Tunari in Cochabamba provided for tariff indexation. The study showed that, over the period 1997–1998, tariff adjustment to the US dollar would have resulted in an 11.5% tariff increase due to currency devaluation. The Cochabamba concession agreement was even more favourable to the private operator as tariffs were not only pegged to the US dollar but also to US inflation (Laurie & Crespo, 2002).

This can be contrasted with an approach to currency risk in public water operations (Hall *et al.*, 2002b; Lobina, 2001). As public undertakings are free from the imperative to repatriate dividends in hard currency that affects TNCs' subsidiaries, they face less constraint in tapping loans denominated in local currency.

Renegotiation and Dynamic Pricing

Continuous renegotiation of contractual arrangements in a context of unequal resources and capacity has a direct impact on water pricing, as tariffs may be raised and projected investments revised downwards. Water TNCs often start to renegotiate immediately after the beginning of operations in order to alter the legal and economic framework of operations. The disparity of resources and capacity between international private operators and local authorities almost systematically translates into a penalizing outcome for local consumers and taxpayers, with a net increase in the cost of outputs.

- *Buenos Aires, Argentina:* Having started only eight months after the beginning of operations, continued renegotiation of the Aguas Argentinas concession (Azpiazu & Forcinito, 2002, pp. 23–38) led to a far greater increase in the relative cost of outputs than suggested by nominal increases in tariffs and billing. Figure 3 shows the development of average bills in Buenos Aires from May 1993 to January 2002, as opposed to the variation in Argentine inflation over the same period. Average household bills increased by 88.2% in nominal terms as opposed to a 7.3% increase in the Consumer Price Index (Azpiazu & Forcinito, 2002, pp. 38–39, 55). In order to evaluate the net increase in the cost of outputs under Aguas Argentinas, the reduction of projected investments resulting of renegotiation should also be taken into account. From May 1993 to December 1998, Aguas Argentinas failed to realize 57.9% of the originally agreed investments for a total of US$746.39 million (see Table 1). Table 3 shows that, from May 1993 to December 1998, Aguas Argentinas failed to realize 46.3% of originally agreed investments in the expansion of the water supply network and 56.8% of originally

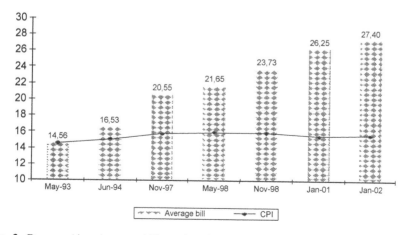

Figure 3. Buenos Aires:Average bills under Aguas Argentinas in Pesos/US$ and inflation, 1993–2002. *Source:* Azpiazu & Forcinito (2002, p. 39).

Table 3. Water and sewerage services: connected population, projected and effective, during the first five-year period (extended, May 1993 to December 1998) (in thousands and percentage)

	Water	Sewerage
According to original bid	1709	924
According to Resolution Etoss No. 81/94	1764	925
According to Decree No. 1,167/97	1504	809
Service expansion		
I. Works by AASA	631	112
II. OPCT*	286	287
III. Regularization of illegal users	172	152
Real expansion of the network (I + II)	917	399
Degree of effective compliance		
(not considering regularization		
of illegal users)		
With respect to the original bid	53.7%	43.2%
With respect to Resolution Etoss No. 81/94	52.0%	43.1%
With respect to Decree No. 1,167/97	61.0%	40.3%

Note: *OPCT: Works on account of a third party paid by the users.
Source: Azpiazu & Forcinito (2002, pp. 42–43); elaboration on the basis of the History and Balances of Aguas Argentinas SA and information from the Users' Committee of ETOSS.

agreed investments in the expansion of the sewerage network. When considering investment targets as defined after renegotiation in 1997, Aguas Argentinas failed to realize 39% of projected expansions in the water supply network and 59.7% of projected investments in the expansion of the sewerage network (Azpiazu & Forcinito, 2002, pp. 39–45).

● *Manila, Philippines:* If Maynilad opened the renegotiation process in December 2000, Manila Water Company started requesting revisions to the agreement in 1998 or less than a year since beginning operations. Manila Water Company sought the modification of originally agreed parameters which would result in an increase in prices above the levels projected by two of its competitors in the bidding process. This implies that, had Manila Water Company put forward its bid on the basis of the parameters it later sought to obtain through renegotiation, it would have failed to win any of the two Manila concessions (Esguerra, 2002, pp. 12–14).

PSP and Delivery to the Poor

The proponents of PSP expect private water operators to deliver better quality services at lower costs in the light of their efficiency and benefit the totality of consumers, including the poor. As summarized by the World Bank, quoted by Catley-Carlson (2002, p. 12): "The reality is that the private sector has the capacity and the interest to serve the poor, is willing to experiment with low cost options, and different levels of service, and with greater efficiency can benefit all consumers".

A critical analysis of the arguments advanced in favour of PSP as an optimal response to the need for serving the urban poor has been elaborated in Hall & Lobina (forthcoming). Such analysis centres on the limits of business in delivering the expected benefits to the poor. Hall & Lobina (forthcoming) conclude that:

- "expectations of the private sector were flawed by failing to anticipate the impact of commercial strategies on determining the boundaries of expansion, the economics of service, and exit strategies;
- the community-based techniques used by the private companies in extending water services to the poor are not innovations but borrowed from public sector and community initiatives;
- the requirements of profit-maximization and risk reduction strategies of the private companies define the scope of private expansion at a narrower and less sustainable boundary than the public sector".

Limitations of Private Sector in Delivering for the Poor and Development

Empirical evidence suggests that the profit motive may be extremely difficult to reconcile with service delivery to the poor without public subsidies or the alteration of the original contractual agreement. In this sense, it is the experience of the major water TNCs that is most illuminating.

In January 2002, in a presentation to the World Bank water division, J.F. Talbot, the chief executive of SAUR International, the fourth largest water company in the world, addressed the limitations of the private sector in delivering for the poor. Talbot (2002) referred to the huge scale of the investments required to meet the basic water needs of developing countries and thus enhance sustainable development, while questioning the commercial viability of operations. Talbot rejected the assumption that privatization would automatically tap into private funds. Moreover, he acknowledged that "The scale of the need far out-reaches the financial and risk taking capacities of the private sector" and finally rejected the sustainability of introducing full cost recovery from users in developing countries: "Water pays for water is no longer realistic in developing countries"; "Service users can't pay for the level of investments required, nor for social projects" (Talbot, 2002, p. 20).

The above suggests that the fundamental problem faced by private operators in trying to provide water supply and sanitation to the poor is that the poor are not profitable, because they cannot afford to pay for the connection or to consume enough water to cover the costs of service provision. When it comes to serving the poor, water TNCs might be confronted with a dilemma originating out of the conflict between the objectives to maximize return for the private operator's shareholders and to ensure universal access to basic water supply and sanitation services. As profit maximization is the 'raison d'être' of private enterprise, the dilemma is between withdrawing from serving areas where consumers cannot afford to provide the operator with the return demanded by its shareholders, or where commercial risk is perceived as excessive, and supplying all customers in the area of operation while sticking to commercial considerations, which in turn inform pricing policy and levels of service and investment. Different approaches to solving this dilemma all have significant implications in terms of access to service. In particular, the operator's choice to keep to commercially desirable tariff levels might lead to what can be defined as 'water poverty', in the case of consumers who cannot afford to pay bills and have to resort to alternative sources of drinking water and face serious health hazards. The following are examples of 'water poverty'.

- *Paraná, Brazil:* Some poor users in the area served by Veolia's subsidiary Sanepar have been forced to resort to alternative sources of water to treated drinking water supplied by the semi-privatized operator. In November 2002, months after being disconnected for failure to pay water bills, poor families in Maringá were using rain water for cooking and drinking purposes.[53] In January 2003, consumers in Vila Democracia, in the metropolitan region of Paraná's capital city Curitiba, were using contaminated water as they could not afford paying for the bills issued by Sanepar.[54]
- *KwaZulu-Natal, South Africa:* In August 2000, a cholera outbreak began in KwaZulu-Natal and by February 2002 the total death toll had reached 260, the worst epidemic in the history of South Africa.[55] The development of the epidemic was associated with the operation of government policies of full cost recovery for water.[56] According to the South African Press Association: "A government committee has found that it would be feasible to provide water free of charge to poor communities, Water and Forestry Minister Ronnie Kasrils said on Friday [October 2000]". Kasrils said health problems, such as the current cholera outbreak in KwaZulu-Natal, arose when the poor were excluded from water supplies. He said his visits to rural areas had highlighted the fact that many people were so desperately poor that they could not afford what might seem to ordinary people a very small price for water. They therefore chose to buy food instead of water and took their chances in searching for river or groundwater.[57] The South African government's approach to providing free water to those who could not afford to pay appeared to reverse the previous policy based on World Bank advice. World Bank economist John Roome advised the then water minister Kader Asmal in 1995 against cross-subsidization, arguing that this could deter private companies (Hall *et al.* 2002a).

The last example shows that when public sector water operators try and introduce commercial policies, such as full cost recovery in pricing, the impacts on the poor in terms of restricted access are not dissimilar to those caused by the commercialization of service provision promoted by the advocates of PSP and adopted by water TNCs.

TNCs' Attempts at Solving Problems with Service Provision to the Poor

In an attempt to escape the dilemma between withdrawing from potential markets and the risk of inducing 'water poverty' in the served areas, water TNCs have developed different approaches aiming at reconciling commercial considerations and the objective of serving the poor. These approaches range from defining the contractual framework in a way to ring-fence profitable customers, to introducing cross-subsidization of tariffs and requiring substantial public sector subsidies and other forms of financial support to private sector operations. So far, empirical evidence on experiences in this sense has highlighted persistent difficulties in resolving the conflicts between private water operators' profit seeking and the pursuit of universal access, leading to question whether PSP can ever be made pro-poor.

Global selectivity. The most basic strategy available to the private companies is avoiding cities and countries which are assessed as not profitable, a selective approach not available to public authorities, who have to address the needs of their territories. This negative selection process is clearly presented in the policy announced by Suez in January 2003,[58] under which the company planned to adopt more demanding investment criteria for investments in developing countries. These criteria gave preference for "currency risk-exempt financing", "the quickest free cash flow generating projects and contracts", and projects which finance all their investments out of their own cash-flow, which will limit investments to populations which can afford prices based on full cost recovery, including the cost of investments.

Ring-fencing profitable customers. In February 2001, a presentation by two Veolia executives at a conference on the reform of the water sector in Africa emphasized the ring-fencing of profitable customers as a method to ensure the viability of PPPs in the region (Bourbigot & Picaud, 2001). More precisely, the presentation stressed that the requirements of low risk and profitability limited private investment to "big cities where the GDP/capita is not too low". In addition, the prospects of profit depended either on "Sufficient and assured revenues from the users of the service", thus potentially excluding the poor and anyone who could not afford to pay for the required stream of revenues, or on government guarantees of payments for the service, in effect subsidies.

Water TNCs have developed a number of complementary approaches. One approach is to redefine the boundaries of the service to be provided. Therefore, in Cartagena, Colombia, for example, the shanty town areas are treated as not covered by the contract because they are not in the city area. In La Paz, Bolivia, where the contract said unequivocally that 100%, including the major shanty town of El Alto had to be connected, Suez subsidiary Aguas de Illimani argues that 'connection' does not mean a piped connection but may just mean access to a standpipe or tanker (Hall, 2002). Similarly, it remains unclear whether the Cordoba, Argentina concession requires the private operator to connect residents in low-income areas to the network (Nickson, 2001b, pp. 21–22).

Financing Water for the Poor and Subsidies for TNCs: The Camdessus Report

The original arguments in favour of water privatization included a criticism of the public sector for failing to charge enough for full cost recovery, leading to excessive and non-transparent subsidization (Winpenny, 1994, p. 5; Idelovitch & Ringskog, 1995, p. 1). Interestingly, subsidies through public finance are now seen as a key to sustain the presence of the private sector in developing countries.

According to SAUR's CEO Talbot, the problems of scale, risk, excessive regulation, and impossibility of full cost recovery, which undermine the attractiveness of the water business for the private sector, can be solved by injections of public sector subsidies. He stated that "substantial grants and soft loans are unavoidable to meet required investment levels" (Talbot, 2002, p. 23) and concluded that, without financial support in the form of subsidies and soft loans under the co-ordination of the World Bank, water TNCs would pull out of developing countries (Talbot, 2002, p. 30).

The so-called Camdessus report on financing water for the poor can be seen as a response to this kind of demand from the water companies. Written by James Winpenny, the report is the contribution of the World Panel on Financing Water Infrastructure, chaired by former Managing Director of the International Monetary Fund (IMF) Michel Camdessus. The report addresses the question of how to finance water infrastructure stemming from the ambitious United Nations Millennium Development Goals (MDGs): to reduce by half the proportion of people, respectively without sustainable access to safe drinking water and without access to basic sanitation, by 2015 (Camdessus, 2003, p. ii).

The devices proposed by the Camdessus report include the use of aid and IFI money to help provide guarantees against political risks, finance private sector tendering costs, and reduce currency risk. All of these devices operate as risk-reducing subsidies to private sector involvement. The guarantees would come from export credit agencies, but most of all from the development banks (Camdessus, 2003, p. 26). The tendering costs would be assisted through the use of aid money in the form of a Revolving Fund or funds that local public authorities would have to replenish "partly or totally . . . on the award of the project to the successful bidder" (Camdessus, 2003, p. 22). The currency risks will be reduced through use of public sector funds to create a "devaluation liquidity backstopping facility" (Camdessus, 2003, p. 29).

Part 3. The Policy of Water Reform: Multilateral Agencies, PSP and the Public Sector Option

Policy Issues

Previous sections of this paper have focused on the developmental impact of PSP in water supply and sanitation. This section discusses some key policy issues arising, focusing on three features in particular:

- the 'policy bottleneck' of IFIs, especially the World Bank;
- the specific problem of conditionality;
- the need to recognize and support the actual and potential role of the public sector.

Multilateral Agencies, PSP and the Policy Bottleneck

Multilateral development agencies are a major driver of water privatization worldwide. Their clout in doing so is commensurate to the resources they can devote to influencing national policies in transition and developing countries through a range of promotion and advisory activities, but most importantly to the strategic role they play in channelling aid and investment finance to recipient countries. In promoting PSP in the water sector, multilateral agencies appear inspired by the theoretical benefits expected from PSP and have generally proved resistant to acknowledge the limitations of the private sector in fostering sustainable water development. Instead of opening up in favour of a wider range of policy options competing on the basis of their responsiveness to local developmental requirements, they have insisted on the promotion of a one-size-fits-all model, that of PSP. The literature promoted by the World Bank's Rapid Response Unit is a good example of this.[59] This results in a 'policy bottleneck', whereby PSP is increasingly relied upon as the way forward irrespective of the adverse implications it has on sustainable water development and macro-economic policies.

The Problem of Conditionality

In the water sector, privatization conditionality has been rejected by the inter-governmental International Conference on Freshwater in Bonn in December 2001 where the delegates declared in their 'Recommendations For Action' that "Private sector participation should not be imposed on developing countries as a conditionality for funding" (Federal Government of Germany, 2001, p. 30).[60] The practice is widespread in poor countries (Grusky, 2001; Hall & de la Motte, 2004).

The World Bank's reliance on the use of conditionality is clear in policy documents. For example, the Private Sector Development (PSD) Strategy that was approved in April 2002 (World Bank, 2002, pp. 49–50, 55), makes it clear that the Bank will continue to force countries to privatize through the use of 'policy-based lending',[61] a clear euphemism for conditionality. One clear example was in Tanzania, where a consortium of Biwater and German engineering firm Gauff was awarded a 10-year water supply and sanitation lease contract for Dar es Salaam in January 2003. The operations, due to start in June 2003, would be financed by a total US$145 million package coming from a number of multilateral and bilateral agencies: the African Development Bank (AfDB), the World Bank, EIB and the Agence Française de Développement (AFD). According to the Global Water Report: "All four bodies insisted on a private concession as a precondition for committing funds".[62]

Need for Strengthening the Public Sector Option

Global policy urgently requires a reinstatement of public sector water operations at the heart of future strategies, especially given the political commitment to the ambitious MDGs[63] and the parallel retreat of water TNCs from developing markets perceived as excessively risky or unprofitable (Hall & Lobina, 2004, pp. 269–270). The simple fact is that "over 90% of domestic water and wastewater services worldwide are provided by the public sector and this is likely to remain the case" (Rogers & Hall, 2003, p. 32).

In this sense, there are three key areas where to converge the efforts of the international water community:

1. Analyse existing successes of public sector. While the World Bank has repeatedly emphasized the problems experienced with public sector operations, very little effort has been made to examine successful public sector operations and the lessons that can be learnt. Case studies such as those of Porto Alegre's municipal water company DMAE (Hall *et al.*, 2002b) show the positive developmental impact produced by sound management practices together with transparency, accountability and the democratization of management through meaningful public participation. Other case studies corroborate the potential for publicly-owned and managed water operations in transition and developing countries (Lobina & Hall, 2000; Hall, 2001).
2. Building capacity. In many cases there may be a need for capacity-building to enable public operations deliver the required service. There are good established examples of public-public partnerships (PUPs) which have done this, especially in the Baltic region, which can be used as a basis for developing a more general policy (Hall & Lobina, 2003). International organizations of public sector water companies also have a key role to play.

3. Developing finance mechanisms for local public sector. The financial focus needs to be switched from providing incentives and subsidies for multinational operators, to identifying the financial needs of local public sector water providers, and ways in which they can be supported. Extension of work on municipal bonds, and revolving funds for municipalities to draw on, is a key part of this.

In the light of political resistance to privatization, "seen as benefiting mainly elite and corrupt interests at home and abroad" and the withdrawal of water TNCs from developing countries, expectations on the merits of PSP have declined considerably (Hall & Lobina, 2004, pp. 266–267). The World Bank has recently acknowledged that "there was probably some 'irrational exuberance' in recent years on the potential benefits of privatization" (CEFE (Competency based Economies through Formation of Enterprise), 2004, p. 1).

Conversely, the April 2004 Operational Guidance for World Bank staff state that "The Bank will work with well-performing publicly owned and -operated utilities as well as those that put in place a credible program to improve performance over time" (World Bank, 2004, p. 14). Nevertheless, in view of the extent to which their policies have been skewed in favour of PSP, it remains to be seen whether the World Bank and other IFIs will successfully act as catalysts for the full developmental potential of public water undertakings.

Notes

1. See also http://lnweb18.worldbank.org/eap/eap.nsf/Attachments/water3
2. French transnational corporations in particular have used various names for both the groups and the water sections since 1990. Throughout this paper, for the sake of clarity, the two largest groups will be referred to as 'Suez' and 'Veolia', even when referring to dates before these names were formally used. The principle names associated with these groups are: Suez = Lyonnaise des Eaux; Ondeo. Veolia = Générale des Eaux; Vivendi.
3. Source: PSIRU database; Tim Webb, Vivendi weighs up AWG's £270m arm, *The Business*, 15 December 2002, p. 4.
4. Source: PSIRU database, News Item 3717, Azurix and Wessex face problems of failing to capture water market. Available at http://www.psiru.org/newsitem.asp?newsid = 3717; Jenalia Moreno, Azurix testing the waters—Company competing against giant foes in Latin America market, *The Houston Chronicle*, 9 January 2000.
5. Source: PSIRU database, News Item 4481, Azurix to be broken up. Available at www.psiru.org/newsitem.asp?newsid = 4481; Sophie Barker, Wessex Water parent plans major surgery, *The Daily Telegraph*, 19 April 2001; Micheal Davis, Water under the bridge; Enron will take Azurix private, sell its assets, *The Houston Chronicle*, 16 December 2000.
6. Source: PSIRU database, News Item 4416, RWE to take over Thames Water. Available at http://www.psiru.org/newsitem.asp?newsid = 4116; Andrew Taylor & Uta Harnischfeger, RWE acquires Thames Water in £4.3bn deal, *The Financial Times*, 22 September 2000.
7. Source: PSIRU database; IDB, Aguas del Gran Buenos Aires Water and Sanitation Project, *Project Summary*. Available at http://www.iadb.org/exr/doc98/pro/aar0282.pdf; OSBA receives 1 bid for 6th sub-region—Argentina, *Business News Americas*, 8 July 1999.
8. Source: PSIRU database; Thames wins ESSAM, no bids for ESSAR – Chile, *Business News Americas*, 13 November 2001; Thames Water, Thames Water announces new Chilean concession, *Press Releases*, 13 November 2001. Available at http://www.thameswater.co.uk.
9. Source: PSIRU database; IDB, IDB Approves $40 Million to Improve Water, Sanitation in Guayaquil—New resources for system through private sector concession, Press Release 164/97, 16 July 1997. Available at http://www.iadb.org/exr/PRENSA/1997/cp16497e.htm; International Briefs, *Global Water Report*, 26 September 1997; International Water Services, Guayaquil concession confirmed—Ecuador, *Business News Americas*, 8 January 2001. See also Acosta (2002).

10. Source: Available at http://www.legifrance.gouv.fr/WAspad/UnTexteDeJorf?numjo = PRMX9200148L.
11. Source: Available at http://www.legifrance.gouv.fr/WAspad/UnTexteDeJorf?numjo = ENVX9400049L.
12. Nice work cuts costs, Global Water Report 144, 19 April 2002, p. 3.
13. Informe De Pricewaterhousecoopers Valencia Pagara A Avsa 54 Millones Si Rescata La Concesion, *Expansion*, 17 January 2001.
14. Alfonso, J. (2001) Aguas de Valencia gana el concurso de suministro local, *Cinco Dias*, 3 October 2001, p. 6.
15. Enron preferred bidder for Mendoza: Argentina/Project, Global Water Report FT Bus Rep: Energy, 26 May 1998.
16. Source: PSIRU database; Iskandar, S. & Taylor, A. (1999) Franco-Spanish group wins bid for Emos, *Financial Times*, 15 June, p. 36.
17. Source: PSIRU database; CHENGDU Generale des Eaux-Marubeni Waterworks Company Limited, ADB Private Sector Project Profiles. Available at www.adb.org/Work/Projects/Profiles.
18. Source: PSIRU database; Lippens, G. & Dang, K. M. (2001) Market maker; Asia Focus—Vietnam, *Project Finance*, 1 December, p. 41.
19. Source: PSIRU database; Demand shortfall triggers Eurawasser price rises in Rostock, *FT Energy Newsletters—Water Briefing*, 22 February 1995, p. 11.
20. Source: PSIRU database; Bernard Simon, Municipal partnership pioneer in a squeeze, *Business Day*, 6 June 2001.
21. Source: PSIRU database; Government of Belize, Privatization of WASA—Joint Media Release, *Press Release*, 23 March 2001. Available at http://www.belize.gov.bz/features/wasa/ministerstatement.html.
22. Source: PSIRU database; Cutlack, M. When the dollars run out, *New Statesman*, 4 March 2002.
23. Source: PSIRU database; Interaguas Allocates US$144mn for Campo Grande—Brazil, *Business News Americas*, 20 July 2000; Guariroba to Invest US $ 7.6mn in 2002—Brazil, *Business News Americas*, 5 November 2001.
24. NRW stands for Non-Revenue Water and is a synonym for UFW (Unaccounted-For-Water).
25. Source: PSIRU database; Bernard Simon, Municipal partnership pioneer in a squeeze, *Business Day*, 6 June 2001.
26. Source: Available at http://www.worldbank.org/icsid/cases/pending.htm.
27. Source: PSIRU database; Hudson, P. Muddy waters—Overview of troubles with Argentina's water infrastructure, Latin Trade Business & Industry: 5 March 1999.
28. See also: Argentina companies must settle dispute with govt locally before arbitration -report, *AFX European Focus*, 5 July 2002.
29. Source: PSIRU database; Aguas del Tunari resorts to World Bank's ICSID arbitration and claims US$25 million compensation, News Item 4983. Available at http://www.psiru.org/newsitem.asp?newsid = 4983.
30. Sources: PSIRU database; Buenos Aires government asks for Azurix pledge to improve water service, News ID 4439; Azurix provides low pressure water in Buenos Aires, News ID 3867; Buenos Aires Guarantees Water Supplies—Argentina, *Business News Americas*, 5 November 2001; Available at http://www.worldbank.org/icsid/cases/pending.htm; SOSBA (Sindicato de Obras Sanitarias); Empresa estadounidense de agua Azurix continuara prestando servicio en Argentina, *Agence France Presse*—Spanish, 1 January 2002; BA to file US$600mn claim against Enron, *Business News Americas*, 12 August 2002; Argentine Province Sues Enron, *The Oil Daily*, 13 August 2002.
31. Sources: Potsdam Management Award: Germany/Privatization, *Global Water Report*, 8 January 1998; GERMANY/Companies—Vivendi spells out its strategy, *Global Water Report*, 10 November 2000; *Berliner Morgenpost*, 27 July 2000; Potsdam City Council, Press Release No. 42/01, 31 January 2001.
32. Source: PSIRU database; News Item 4777, WSSA South African water contract nullified. Available at http://www.psiru.org/newsitem.asp?newsid = 4777.
33. Source: PSIRU database; Vivendi Water reaches agreement with Szeged council, *MTI Econews*, 12 February 2001.
34. Source: Democracy Centre, Cochabamba, Bolivia.
35. Agreement for Management, Operation and Maintenance of the Water and Sewage Systems of Fort Beaufort and Associated Customer Management, 5 October 1995, p. 6.
36. NEPSZABADSAG: 7 December 1998.
37. Source: *Le Monde*, 28 January 1997.
38. Issues paper, NIS Ministerial Conference on water management and Investment, Almaty 16–17 October 2000.
39. Source: PSIRU database; Cecille S. Visto & Ruffy L. Villanueva, Rate hike to boost water services, *BusinessWorld*, 12 November 2002, p. 1.

40. Source: PSIRU database; Ruffy L. Villanueva, MWSS-RO readying higher water tariffs, *BusinessWorld*, 14 November 2002, p. 1.
41. Jugement de la 7°chambre de la Cour d'Appel de Lyon du 9 juillet 1996; Cour de cassation chambre criminelle du 27 octobre 1997. No 96-83.698.
42. Chambre régionale des Comptes Rhône-Alpes, *Observations définitives de la gestion des services de l'eau et de l'assainissement de la commune de Grenoble (Isère)*: 24 novembre 1995, p. 21.
43. Source: PSIRU database; News Item 4777, WSSA South African water contract nullified. Available at http://www.psiru.org/newsitem.asp?newsid = 4777.
44. Source: PSIRU database; Private operators to refinance sell-off costs—Paraguay, *Business News Americas*, 28 July 2000; WB delays clearance of Corposana rules—Paraguay, *Business News Americas*, 13 November 2000.
45. Source: PSIRU database; Holly Williams, Sanchis scoops Paraguay privatizations, *PR Week*, 15 June 2001; Sanchis & Asociados informe sobre privatizaciones en Paraguay, *Expansion*, 5 June 2001.
46. Source: PSIRU database; IDB, IDB Approves $40 Million to Improve Water, Sanitation in Guayaquil—New resources for system through private sector concession, Press Release 164/97, 16 July 1997. Available at http://www.iadb.org/exr/PRENSA/1997/cp16497e.htm; International Briefs, *Global Water Report*, 26 September 1997; International Water Services, Guayaquil concession confirmed—Ecuador, *Business News Americas*, 8 January 2001. See also Acosta (2002).
47. Source: PSIRU database; France/Companies—Suez results, *Global Water Report*, N. 130, 2 October 2001, p. 8.
48. Source: PSIRU database; France/Companies—Suez AGM, *Global Water Report*, N. 145, 3 May 2002, p. 8.
49. Source: PSIRU database; France/Companies—Vivendi share doubts, *Global Water Report*, N. 127, 3 August 2001, p. 5.
50. Source: PSIRU database; Anglian Water takes controlling interest in Chile water concession, News Items, No. 3384. Available at http://www.psiru.org/newsitem.asp?newsid = 3384; Esval awards contract to Anglian Water's Purac, News Items, No. 4443. Available at http://www.psiru.org/newsitem.asp?newsid = 4443.
51. Source: PSIRU database; Degremont awarded major contract by EMOS, News Items, No. 4427. Available at http://www.psiru.org/newsitem.asp?newsid = 4427.
52. Source: PSIRU database; Aguas Argentinas, water pricing, contractual flexibility and the Argentine crisis, *News Items*, No. 4995. Available at http://www.psiru.org/newsitem.asp?newsid = 4995.
53. Source: Observatório Social; Famílias usam água de chuva, *Gazeta do Povo* (PR), 19 November 2002.
54. Source: Observatório Social; Consumidores da região metropolitana de Curitiba não contam com tarifa social de água, *Paraná Online* (PR), 08 January 2003.
55. Sources: No.5083: *Business Day* 17 November 1999, Will the World Bank halt corruption and cancel debt? Patrick Bond and David Letsie. No.5084: SAPA 26/09/2000, KZN cholera outbreak spreads, death toll rises. No.5085: *Mail* and *Guardian* 13 October 2000, More government confusion on privatization?. No.5086: SAPA 13/10/2000, FREE WATER IS FEASIBLE: KASRILS. No.5087: SAMWU 17/10/2000, Cholera epidemic would not have happened if there was free water! No.5090: Mizanet/IPS 18/10/2000, Govt promises free water to curb cholera epidemic. Available at http://www.iclinic.co.za/oct00/cholera18b.htm/LS. No.5111: *The Lancet* 27/01/2001, Prevention fails to halt South Africa's well-treated cholera epidemic. No.5112: Beeld 23/01/2001, 23/01/2001. No.5113: *The Star* 23/01/2001, Squatters on 'cholera river' to be moved. No.5114: SAMWU press statement 23/01/2001, Union highly concerned about events in Johannesburg, including cholera in Jukskei River. No.5115: *Daily Dispatch* 19/01/2001, Cholera found in Alexandra's Jukskei River. No.5116: Sapa 15/01/2001 GOVT RUNNING LOW ON FUNDS: PEOPLE MUST BUILD THEIR OWN TOILETS. No.5117: *Business Day* 11/01/2001, Cholera cases now more than 17000. No.5118: *BuaNews* 06/11/2000, Safe water and sanitation still elude cholera-plagued KZN. No.5119: *Financial Mail* 03/11/2000, LEARNING TO PADDLE UPSTREAM. No.5120: SAMWU and RDSN 26/10/2000, The Cholera Epidemic in South Africa and the Government's attack on the Working Class. No.5121: Beeld 25/10/2000, Supply cuts caused cholera. No.5122: Sapa 29/01/2001, Cholera death toll hits 85.
56. *Mail* and *Guardian* 13 October 2000.
57. Free water is feasible, 13 October 2000, Sapa.
58. SUEZ introduces its 2003–2004 action plan: refocus, reduce debt, increase profitability, Suez Press Releases, 9 January 2003. Available at http://www.suez.com/upload/up970.pdf.
59. See http://rru.worldbank.org/.
60. See also http://www.water-2001.de/outcome/BonnRecommendations/Bonn_Recommendations.pdf.

61. See the Implementation Matrix, Annex I, Private Sector Development Strategy—Directions for the World Bank Group, 9 April 2002. Available at http://rru.worldbank.org/documents/PSDStrategy-April%209.pdf.

62. Biwater wins Dar es Salaam concession, *Global Water Report* 162, 15 January 2003, pp. 1–2.

63. See Camdessus (2003), p. ii.

References

Acosta, J. (2002) Ecuador/workers rise up against privatization in Guayaquil, *Defend the Global Commons*, 1(1), pp. 18–19, FebruaryAvailable at http://www.citizen.org/documents/ACFC0.pdf

ADB (2000) *Developing Best Practices for Promoting Private Sector Investment in Infrastructure—Water Supply* (Manila, Asian Development Bank). Available at (http://www.adb.org/Documents/Books/Developing_Best_Practices/Water_Supply/water_supply.pdf).

ADB (2004) *Water in Asian Cities — Utilities' Performance and Civil Society Views* (Manila: Asian Development Bank). Available at: http://www.adb.org/Documents/Books/Water_for_All_Series/Water_Asian_Cities/default.asp#contents.

Azpiazu, D. & Forcinito, K. (2002) Privatization of the water and sanitation systems in the Buenos Aires Metropolitan Area: regulatory discontinuity, corporate non-performance, extraordinary profits and distributive inequality. Paper presented at the First PRINWASS Project Workshop, University of Oxford, 22–23 April. Available at http://www.geog.ox.ac.uk ~ prinwass/Azpiazu_Forcinito.PDF

Bakker, K. & Hemson, D. (2000) Privatizing water—BOTT and hydropolitics in the new South Africa, *South African Geographical Journal*, 82(1), pp. 3–12.

Bayliss, K. (2001). Water privatization in Africa: lessons from three case studies, *PSIRU Reports*, May. Available at http://www.psiru.org/reports/2001-05-W-Africases.doc

Bourbigot, M.M. & Picaud, Y. (2001) Public-Private Partnership (PPP) for municipal water services. Vivendi Water presentation at Regional Conference on The Reform of the Water Supply and Sanitation Sector in Africa, Enhancing Public-Private Partnership in the Context of the Africa Vision for Water (2025), Kampala, Uganda, 26–28 February. Available at http://www.wsp.org/english/afr/wup_conf/v2_municipal.pdf

Braadbaart, O. (2001) Privatizing water. The Jakarta concession and the limits of contract. Paper presented at KITLV Jubilee Workshop on Water as a Life-giving and a Deadly Force, Leiden, the Netherlands, 14–16 June.

Camdessus, M. (2003) *Financing Water for All—Report of the World Panel on Financing Water Infrastructure.* Chaired by Michel Camdessus. Report written by James Winpenny. Presented at 3rd World Water Forum, Kyoto, Japan, 16–23 March 2003. Available at http://www.worldwatercouncil.org/download/CamdessusReport.pdf.

Catley-Carlson, M. (2002) Spans and suspensions: building bridges and water security through integrated water resource management, *Water Science and Technology*, 45(8), pp. 7–17, (*Proceedings of the 11th Stockholm Water Symposium, Water Security for the 21st Century—Building Bridges through Dialogue, Stockholm, Sweden, 13–16 August 2001* (London: IWA Publishing).

CEFE (2004) 'Credible regulation vital for infrastructure reform to reduce poverty, says World Bank'. Press release, 14 June (Bad Homburg v.d.H., Germany, Competency based Economies through formation of Enterprise). Available at http://www.cefe.net/forum/CredibleRegulation.pdf

Conseil de la Concurrence (2002) Décision n° 02-D-44 du 11 juillet 2002 relative à la situation de la concurrence dans les secteurs de l'eau potable et de l'assainissement, notamment en ce qui concerne la mise en commun des moyens pour répondre à des appels à concurrence. Available at http://www.finances.gouv.fr/reglementation/avis/conseilconcurrence/02d44.htm

Cour des Comptes (1997) La gestion des services publics locaux d'eau et d'assainissement, Les éditions du *Journal*, Paris.

Druin, O. (2002) Les règles de la concurrence sont contournées, in: le scandale du prix de l'eau, *Capital*, October pp. 116–118.

Esguerra, J. (2002) The Corporate Muddle of Manila's Water Concessions: How the world's biggest and most successful privatization turned into a failure. Study commissioned by WaterAid, London. Available at http://www.ipd.ph/pub/wip/IPDWIP articleonmwss.pdf

Federal Government of Germany (2001) Water—A Key to Sustainable Development, Conference Report: International Freshwater Conference, Bonn, 3–7 December. Available at http://www.water-2001.de/ConferenceReport.pdf

Franceys, R. (2000). Water and public-private partnerships. Keynote speech, special subject on Water and Public-Private Partnerships, 2nd World Water Forum, The Hague, The Netherlands, 17–22 March.

Gaillard, E. (2003) International Arbitration Law—'Vivendi' and Bilateral Investment Treaty Arbitration, *New York Law Journal*, 229, 6 February, p. 3.

Grusky, S. (2001) Privatization tidal wave: IMF/World Bank Water Policies and the price paid by the poor, *Multinational Monitor*, 9(22), 1 September, p. 14.

Hall, D. (1997). Public partnership and private control—ownership, control and regulation in water concessions in central Europe, *PSIRU Reports*, May. Available at http://www.psiru.org/reports/9705-W-Eur-JV.doc

Hall, D. (1998) Restructuring and privatization in the public utilities—Europe, in: L. de Luca (Ed.) *Labour and Social Dimensions of Privatization and Restructuring (Public Utilities: Water, Gas and Electricity)*, pp. 109–151 (Geneva: International Labour Office).

Hall, D. (1999) Privatization, multinationals and corruption, *Development in Practice*, 9(5), November, pp. 539–556, Available at http://www.psiru.org/reports/9909-U-U-Corrup.doc

Hall, D. (2001) Water in public hands—public sector water management, a necessary option, *PSIRU Reports*, June. Available at http://www.psiru.org/reports/2001-06-W-public.doc

Hall, D. (2002a) The water multinationals 2002—financial and other problems, *PSIRU Reports*, August. Available at http://www.psiru.org/reports/2002-08-W-MNCs.doc

Hall, D. (2002b) Secret reports and public concerns—a reply to the USAID paper on Water Privatization 'Skeptics', *PSIRU Reports*, August. Available at http://www.psiru.org/reports/2002-08-W-Skeptics.doc.

Hall, D. & Lobina, E. (2001) Private to public: international lessons of water remunicipalization in Grenoble, France, *PSIRU Reports*, August. Available at http://www.psiru.org/reports/2001-08-W-Grenoble.doc

Hall, D. & Lobina, E. (2002) Water privatization in Latin America, 2002, *PSIRU Reports*, July. Available at http://www.psiru.org/reports/2002-06-W-Latam.doc

Hall, D. & Lobina, E., (2003) International solidarity in water—public-public partnerships in North-East Europe, *PSIRU Reports*, March. Available at http://www.psiru.org/reports/2003-03-W-NEeurope.doc.

Hall, D. & Lobina, E. (2004) Private and public interests in water and energy, *Natural Resources Forum*, 28, pp. 266–275.

Hall, D. & Lobina, E. (forthcoming) Profitability and the poor—corporate strategies, innovation and sustainability, *Geoforum*.

Hall, D. & de la Motte, R. (2004) Dogmatic development: privatization and conditionalities in six countries, *PSIRU report for War on Want*, February. Available at http://www.psiru.org/reports/2004-02-U-condits.pdf; http://www.psiru.org/reports/2004-02-U-condits-ann.doc

Hall, D., Bayliss, K. & Lobina, E. (2002a) Water privatization in Africa, *PSIRU Reports*, June. Available at http://www.psiru.org/reports/2002-06-W-Africa.doc

Hall, D., Lobina, E., Viero, O.M. & Maltz, H., (2002b). Water in Porto Alegre, Brazil—accountable, effective, sustainable and democratic, *PSIRU Reports*, August. Available at http://www.psiru.org/reports/2002-08-W-dmae.pdf

Hall, D., Lobina, E. & de la Motte, R. (2005) Public resistance to privatization in water and energy, *Development in Practice*, 15 (3/4).

Hellman, J., Jones, G. & Kaufmann, D. (2000) Are foreign investors and multinationals engaging in corrupt practices in transition economies?, *Transition*, May–June-July, pp. 4–7, Available at http://www.worldbank.org/wbi/governance/pdf/fdi_trans_0800.pdf

ICSID (2000) Compañía de Aguas del Aconquija, S.A. & Compagnie Générale des Eaux, Claimants v. Argentine Republic, Respondent (Case No. ARB/97/3) Award, *Online Decision and Awards*, 21 November. Available at http://www.worldbank.org/icsid/cases/ada_AwardoftheTribunal.pdf.

Idelovitch, E. & Ringskog, K. (1995) *Private Sector Participation in the Water Supply and Sanitation in Latin America* (Washington DC: The World Bank).

Klein, M. (1996) Economic regulation of water companies, Policy Research Working Paper 1649 (Washington DC: The World Bank), September 1996. Available at http://rru.worldbank.org/documents/economic_regulation_water_companies.pdf.

Klein, M., So, J. & Shin, B. (1996) Transaction Costs in Private Infrastructure Projects—Are They Too High?, *Public Policy for the Private Sector*, Note No. 95, October (Washington DC: The World Bank Group), Available at http://rru.worldbank.org/viewpoint/HTMLNotes/95/95klein.pdf

Laurie, N. Crespo, C. (2002) Contextos cambiantes para el desarrollo de iniciativas en favor de los pobres (pro-poor) a través de concesiones del agua. Casos la Paz/El Alto Cochabamba. Presentation to DFID-sponsored workshop, la Paz, 19 April.

Lobina, E. (2000) Cochabamba—water war, *FOCUS on the Public Services*, 7(2), pp. 5–10, June. Available at http://www.psiru.org/reports/Cochabamba.doc

Lobina, E. (2001) Water privatization and restructuring in Central and Eastern Europe, 2001, *PSIRU Reports*, December. Available at http://www.psiru.org/reports/2001-11-W-CEE.doc

Lobina, E. & Hall, D. (2000) Public sector alternatives to water supply and sewerage privatization: case studies, *International Journal of Water Resources Development*, 16(1), pp. 35–55.

Lorrain, D. (1997) Introduction—the socio-economics of water services: the invisible factors, in: D. Lorrain (Ed.) *Urban Water Management—French Experience Around the World*, pp. 1–30 (Levallois Perret: Hydrocom).

Ménard, C. & Clarke, G. (2000). A transitory regime water supply in Conakry, Guinea, Policy Research Working Paper 2362, June (Washington DC: World Bank, Development Research Group). Available at http://econ. worldbank.org/docs/1116.pdf

Miller, C. (1999) Communities and public-private partnerships—theory and roles related to the provision of water and sanitation, Research Paper (New Haven: UNDP/Yale Collaborative Programme Research Clinic).

Muñoz, A.D. (2002) El agua: derecho humano o mercancia de lucro? —Suez Lyonnaise des Eaux. Paper for Union de Usuarios y Consumidores.

Nickson, A. (1996) Urban water supply sector review. Papers in the Role of Government in Adjusting Economies, 7, January (Birmingham: International Development Department, University of Birmingham, for the Economic and Social Research Council of the British Government).

Nickson, A. (2001a). Establishing and implementing a joint venture—water and sanitation services in Cartagena, Colombia. Working Paper 442 03, January (London: GHK International). Available at http://www.ghkint. com/downloads/Cartagena.pdf

Nickson, A. (2001b). The Córdoba water concession in Argentina. Working Paper 442 05, January (London: GHK International). Available at http://www.ghkint.com/downloads/cordoba.pdf

PSPRU (1998) Łodz, Poland: successful campaign against water privatization 1993–94. Report by the Public Services Privatization Research Unit. Available at http://www.psiru.org/educ/Resources/ Res8.htm.

Rees, J. A. (1998) Regulation and private participation in the water and sanitation sector, *Natural Resources Forum*, 22(2), May (also published as TAC Background Paper No. 1, Stockholm, Global Water Partnership. Available at http://www.gwpforum.org/gwp/library/Tac1.pdf.

Rogers, P. & Hall, A. (2003) Effective water governance. TEC Background Paper No.7, Global Water Partnership Technical Committee (TEC), February. Available at http://www.gwpforum.org/gwp/library/TEC%207.pdf

Suez Lyonnaise des Eaux (1999) *Suez Lyonnaise des Eaux in 1999*. Available at http://www.suez.com/ documents/english/Suez-ra99.pdf

Talbot, J.F. (2002) Is the international water business really a business? World Bank Water and Sanitation Lecture Series, 13 February. Available at http://www.worldbank.org/wbi/B-SPAN/docs/SAUR.pdf.

Webb, M. & Ehrhardt, D. (1998). Improving water services through competition, Public Policy for the Private Sector, Note No. 164, December (Washington DC: The World Bank Group). Available at http://rru. worldbank.org/viewpoint/HTMLNotes/164/164webb.pdf

West, G. T. (1996) Managing project political risk: the role of investment insurance, *The Journal of Project Finance*, Winter, pp. 5–11.

Willner, J. (2001) Ownership, efficiency, and political interference, *European Journal of Political Economy*, 17(4), November pp. 723–748.

Winpenny, J. T. (1994) *Managing Water as an Economic Resource* (London and New York: Routledge).

World Bank (2002) *Private Sector Development Strategy—Directions for the World Bank Group*, 9 April (Washington DC: The World Bank Group), Available at http://rru.worldbank.org/documents/ PSDStrategy-April%209.pdf

World Bank (2004) Public and private sector roles in water supply and sanitation services, Operational Guidance for World Bank Group Staff, April (Washington DC: The World Bank). Available at http://www.worldbank. org/watsan (accessed 18 October 2004).

Key Issues and Experience in US Water Services Privatization

JEFFREY W. JACOBS* & CHARLES W. HOWE**

Water Science and Technology Board, National Research Council, Washington DC, USA, Institute of Behavioral Science, University of Colorado, Boulder, CO, USA

Introduction

The term 'privatization' covers a broad spectrum of water utility management, operations, ownership arrangements, ranging from private provision of services and supplies (e.g. laboratory analysis), to contracting with a private firm for operations and maintenance or, in an option referred to as 'design, build, and operate' (DBO), contracting for plant design and construction. The outright sale of water utility assets to a private firm represents another form of privatization, although no major US city has sold its water system assets in recent decades. A more common expression of privatization in the US has been the outsourcing of operations and maintenance from a public utility to a private firm.

Several factors are driving US municipal officials to consider privatizing their drinking water and wastewater treatment systems. A key factor is a large backlog of deferred maintenance of water storage, treatment and distribution systems in the US. Estimates of the costs for maintaining and upgrading the US water treatment and distribution infrastructure are staggering; some estimates place the figure for investments necessary for infrastructure replacement at $250 billion, or greater, in the next 30 years.

Utilities are also challenged to comply with increasingly stringent water quality standards. The Safe Drinking Water Act of 1974 continues to play an important role in utility management and operations. Water utility plant operations are becoming increasingly complex and it is often difficult to recruit and retain qualified professionals and to keep pace with technological changes.

The contemporary era of water utility management has broadened beyond a focus on technical excellence and now includes co-operative decision making, public participation

Table 1. Community water systems (public and private) in the United States and population served, 1999

Population served	No. of systems	Percentage of water systems	Population served	Percentage of population served
25–500	31 904	59.2	5.2 million	2.0
501–3,300	14 040	26.0	19.8 million	7.8
3301–10 000	4 356	8.1	25.4 million	10.0
10 001–100 000	3 276	6.1	91.0 million	35.9
>100 000	347	0.6	112.4 million	44.3
Total	53 923	100.0	253.8 million	100.0

Sources: EPA (1997, 1999).
Notes: Total systems based on US Environmental Protection Agency, EPA; Drinking Water Information System Factoids: FY1999 Inventory Data; Ownership percentages based on US Environmental Protection Agency, 1995; Community Water System Survey, and applied to factoid data.

and customer relations. There have also been changes in the educational background of water utility leaders and managers. Civil engineers in the field today have generally studied subjects that emphasize general processes and concepts, as opposed to basic public works technologies. There are also more water utility managers today with backgrounds in economics, public administration, and law than in the past.

There are several large water service delivery firms, based in the US and abroad, that possess ample resources, competent business and technical staff, and state-of-the-art laboratory and other facilities, that are seeking to increase their share in the US water utility sector. The larger global firms include Suez (based in France) and its water division, ONDEO, the German-based multi-utility firm RWE AG, and the French firm, Vivendi Environment. US companies include American Water Works Company, Inc. (which merged with RWE AG in 2002), Philadelphia Suburban Corporation and San José Water. The pace of change and mergers among contractors in today's global water market is rapid, and the number and size of these private water companies has changed greatly in the past five years owing to mergers and acquisitions (PWF, 2001).

Privatized water services have a long history in the US, and the first water systems in large US cities were private ventures. As the nation's cities expanded, however, the resources required to adequately maintain and extend the water infrastructure typically were often beyond the means of the private sector. Today, investor-owned water utilities account for about 14% of total US water revenues, a market share that has held remarkably steady since the Second World War (EPA, 1997). There are currently about 54 000 US community water systems, the majority of which serve fewer than 10 000 people (Table 1). These smaller to medium-sized utilities often face the greatest difficulties in meeting infrastructure needs and in satisfying water and wastewater effluent standards.

Concerns over Privatization

Privatization often represents a viable alternative for water utilities facing the challenges of meeting stringent requirements with limited resources and ageing infrastructure. Water services privatization does not, however, represent a panacea for drinking water and wastewater management. Given the complexities surrounding privatization, and the uniqueness of a water utility and its community or city, it is difficult to identify clear trends or universal rules identifying privatization's pros and cons.

The complexities of privatization contracts require that contracts between the public utility and the private contractor be carefully negotiated and structured lest problems, such as unrealistic contract conditions, arise later during the life of the contract. Contentious situations can arise when the terms of a privatization contract are not realistic, and there are instances in which expected benefits of privatization were not realized, or in which miscommunication resulted in dissatisfaction. For example, the city of Indianapolis, Indiana, recently repossessed its water utility from a private contractor, and, in response to concerns over its contract with a private firm, the city of Atlanta appointed an independent committee in 2002 to review its water services contract with a private provider.

Shortfalls in expectations can be resolved through negotiations between the private contractor and the city. For example, in the state of Ohio, Clermont County's water treatment plant suffered problems of discoloured tap water shortly after a private firm assumed plant operations. It was discovered that the problem would have occurred regardless of who was managing the plant, and adjustments to the contract were subsequently made, which resulted in a 5% rate reduction. Such examples point to the importance and the challenge of identifying whether utility management failures (and successes) were caused by a public or private operator, or whether they would have occurred independently of the operator's status (private or public).

Concerns of Communities

Several concerns regarding water services privatization go beyond privitization's implications for water quality and monthly water fees. Community leaders and citizens may be concerned about how channels of communication might affect, for example, the protection of watersheds that serve as source areas of raw water supplies, or public participation in policy decisions.

Community leaders are also concerned about the possible loss of control over a vital public service. Public leaders are naturally reluctant to surrender control of a water utility, as the public official can never fully transfer accountability for water utility management to the private sector. If a privatization arrangement fails to meet the public's expectations, citizens are more likely to lodge protests with the public official rather than the private contractor. Community leaders must also ensure that some recourse is available in the event that a privatization arrangement does not develop as intended, by ensuring that essential skills and equipment can be regained promptly, if necessary. In the end, water privatization represents a net political gain when the costs of ceding control over the utility are more than offset by improvements in water services delivery.

Another community-level concern relates to the degree of competition with privatized water services. There may be a tendency to equate privatization and competition, as some may assume that a privately held firm's operations will be more 'efficient' because of its greater exposure to market forces. In practice, however, competition is limited to the period when competitive bids are being accepted. Once a contract is signed for water utility management, operation, or design, only monitoring and enforcement of the contract terms—not market forces—can guarantee expected levels of performance. Moreover, the competitive bidding processes may be subject to political favouritism.

Another concern relates to openness and transparency of utility policies and practices. Deliberations of public bodies are subject to numerous 'sunshine' provisions that require open meetings and records. When a private firm assumes operations or ownership, business practices may not necessarily be readily shared with the public. Such agreements should thus be included in water privatization contracts. A related issue is how privatization affects the welfare of the water utility workforce. There are often concerns that workers may be exploited or that jobs will be lost to non-resident personnel.

These concerns point to the importance of carefully structured contracts between a public utility and a private firm. The contract preparation process can be costly and time-consuming. Public utilities often do not have the in-house legal, fiscal and engineering expertise necessary to prepare adequate contracts, and may therefore need to contract with external experts. Reviewing multiple bids on a contract can also be an expensive and lengthy process.

Concerns of Private Contractors

Concerns regarding water services privatization represent a two-way street, as private contractors must also weigh several factors when entering into privatization agreements. A key concern of private providers is the high cost of preparing a detailed technical and financial proposal for managing a major utility system. Contractors must consider the probability that the awarding process will be fair to all parties and that a contract will actually be signed. For example, requests for contract proposals may be issued with no intention of entering into a contract (a practice which can be used as a means to gauge public managers' performance or to win concessions from a union). The substantial resources required to enter a bidding process, and the uncertainty of the ultimate decision, can be barriers to private contractors.

Shortly after the First World War, Congress granted an interest-rate subsidy to municipal government bonds. By exempting investors from having to pay income tax on bond interest earnings, the federal government granted local borrowers a 2.5–3% (250–300 'basis points') cost advantage over private investors. This ability of the public sector to issue tax-free bonds leads some contractors to feel they compete on an uneven terms with the public sector. There have been, however, some revisions to US tax and financing regulations to help encourage private-sector participation.

Options for Providing Water Services

Whether a public official or community chooses to manage and deliver water services through the public or private sector, it should be recognized that private organizations do not necessarily operate 'efficiently', nor are they inherently more efficient than public organizations. Likewise, public organizations are not always ideal. Neither public nor private organizations automatically entail effectiveness, and there are examples of well-run and poorly-run organizations in both sectors.

Decisions regarding water services privatization are usually made in response to inadequacies, perceived or otherwise, in a public water utility. Public officials have four broad choices when considering options for improving water utility performance: (1) improve public operations; (2) contract services to the private sector; (3) seek public-private co-operative arrangements; and (4) transfer ownership of the utility's assets to the private sector.

Improve Public Operations

Benchmarking and performance measurement. Benchmarking refers to the process of comparing a utility's overall performance, or select processes, with the performance of similar utilities. It is accomplished through approaches that range from informal comparisons of data to sophisticated econometric analyses. The benchmarking concept is used in many nations around the world. For example, it has been used in the UK, where private water utilities are under regulatory control of the Office of Water Services (OFWAT). There is also a 'Water Utility Benchmarking Association' that conducts benchmarking studies to identify practices that improve the overall operation of its members (see http://www.waterbenchmarking.com, accessed 3 December 2004). The International Organization for Standardization (ISO) has also developed a set of international standards for utility performance. Based in Switzerland, the ISO provides a basis for certifying enterprises based upon performance benchmarks, and certification under ISO 9000 and ISO 14000 uses a standardized system for assessing company performance.

Benchmarking helps public utilities share knowledge and benefits from self-improvement programmes. These efforts can provide an assessment of individual utility performance, and can also be used to evaluate the performance of components of the business, such as quality of services and products, metering and operation and maintenance. Benchmarking does not apply only to the public sector, as both public and private companies engage in these self-improvement practices.

Organizational improvement: 're-engineering.'. The Phoenix Water Services Department (PWSD) began an internal review in 1995 to see how its performance compared with that of well-run public utilities. This review focused on several areas and identified several goals, such as ensuring that no employee involuntarily lost a job, maintaining or improving customer service levels, and emphasizing on-the-job training and cross-training (e.g. encouraging staff to develop skills in various departments throughout the utility). The city of Phoenix was pleased with the efforts, which reportedly produced substantial cost savings and reduced the need to hire additional staff. Successful re-engineering programmes, as demonstrated in Phoenix, will do more than just adjust the organizational chart, but will initiate positive changes in institutional culture.

An important outcome of recent trends in water privatization has been to enhance the performance of public water utilities, much of which has been conducted through the initiation of benchmarking, re-engineering and similar programmes. Although private water firms have not greatly increased their share in the US water market during the 1990s and early 2000s, the spectre of privatization has motivated improved performance of many water utility systems in the US.

Contract Operations to the Private Sector

Most private firms will enter into contractual agreements to operate facilities they do not own. These contractual arrangements are often referred to 'outsourcing'. They cover ancillary services such as meter reading or laboratory services, and are common in the public and private water sectors in the US and in other nations. These arrangements are intended to allow businesses to focus on their core functions and competencies by hiring specialists to perform ancillary duties. Other examples of these contracted duties include design and construction, financing and services related to operation and maintenance such as vehicle maintenance and even public relations.

Governmental bodies usually outsource only a limited portion of water utility operations to the private sector. For example, a 1997 survey of 261 US cities revealed that municipally-owned water systems that operate under a contract with the private sector serve less than 10% of the US population, and that municipally-owned wastewater facilities under contract serve less than 6% of the US population (US Conference of Mayors Urban Water Council, 1997). The most frequently cited reason for outsourcing water or wastewater treatment services was an expected reduction in operating costs (US Conference of Mayors Urban Water Council, 1997; Water Industry Council, 1999). Other factors included political ideology and the challenges of standards compliance. Although this and other types of privatization represent viable options in some circumstances, the private sector's share of drinking water revenues collected in the US today is about 14%, a figure that has held steady since the Second World War.

When a public utility transfers operations to the private sector, the public agency's roles in utility operations often change dramatically. With privatization of operations, local government's role and perspective typically shifts from an emphasis on traditional utility management to an emphasis on contract management and oversight. The talents and skills needed for contract management (e.g. legal, fiscal, performance evaluation) differ greatly than the skills needed for traditional operations (e.g. engineering, public service). Communities considering privatization should recognize the implications of this shift. Public organizations having internal management problems are not likely to be able to effectively manage outside contractors (Scalar, 2000). Successful contract operations depend on good contractor-public agency relations, which are rooted in a contract that clearly states contractor and public agency responsibilities and clear, measurable performance indicators.

Regionalization and small, rural communities. Contract operations are common in smaller, rural communities in the US. Drinking water and wastewater treatment systems in these areas generally serve less than 3300 households and businesses. In the US, these systems comprise 78% of all drinking water systems, with most of them serving fewer than 500 people (Table 1). For such communities, it is often more economical for a single contractor to oversee operations of several small communities, rather than each small community with its own private contractor.

For smaller, remote communities, regionalization—the consolidation of utility management and operations across several communities—has helped achieve economies of scale and performance improvements. The regionalization option can be achieved through both public and private sector arrangements, and there are examples in the US where it has proven to be useful in both circumstances. These regionalization efforts achieve economies of scale through common operations, as well as improved customer service through an organization with greater access to technical and operational skills.

Forms of Public–Private Co-operation

Mixing the roles of the public and private sector is a third option for improving utility performance. This option has been practiced (albeit on a limited scale) in the US water utility sector for many years, and many utilities commonly employ private engineering firms to design new plants, prepare bid specifications and manage new facility construction. The US Environmental Protection Agency, for example, has long advocated 'public-private partnerships' in the water utility sector (EPA, 1990). One form of these partnerships that has been used frequently is the design-build-operate (DBO) option.

The design-build-operate process.   In the DBO process, a corporate entity designs a water or wastewater facility, then builds and operates the plant under a contract, which typically runs for 15 to 25 years. The designer is motivated to anticipate operations problems and to design a plant that will perform efficiently over the contract period. The contractor that provides the proposal chosen by the community is obligated to deliver a constructed facility by a certain date and at a guaranteed cost, and the facility must pass an independent evaluation of its performance. At the end of the contract performance period, the community owns the facility.

Transfer Utility Assets to the Private Sector

Turning over ownership of a water utility's assets to an investor-owned utility is the most extreme form of privatization, and not an option that any major US city has recently exercised, but there are situations in which it represents a reasonable choice (Beecher, 2000). Potential advantages of investor-ownership include local government's release from direct management and planning operations, the transfer of monitoring responsibilities to another body (often a state regulatory agency), the distancing of operations from local political influences, and possible reductions in fraudulent practices. Cities can use the proceeds of the sale for other municipal purposes, and asset sales place the utility under the purview of independent US state economic regulators, who often have a greater capacity for more careful oversight than local governments. Economic regulation requires less duplication of expertise and management than does oversight of a private contract.

There are, however, disadvantages of this option. Accurately valuing utility assets, for example, is a challenging exercise. In the event it is decided to reacquire utility assets, the ensuing process may require a city to exercise powers of eminent domain, and can be costly and controversial. The major barrier, however, is the perceived loss of control of the assets.

Pricing and Regulatory Issues

Pricing

The resources required to maintain and repair an ageing US water delivery infrastructure, and to extend the infrastructure in areas experiencing population growth, are tremendous, and will contribute to increasing costs in the US water industry. Even when stated in comparable dollars, replacement costs far exceed original installation costs. Estimates of the price tag for the needs of the US water and wastewater treatment and distribution infrastructure range from $250 billion in the next 30 years (AWWA, 2001) to roughly $1 trillion in the next 20 years (WIN, 2000).

Many economists agree that water and wastewater services have been historically under-priced. Given the needs of the US water infrastructure, it is becoming increasingly difficult to avoid or postpone the necessary (and often long-neglected) maintenance costs. Limited public funds means that some portion of these costs must be supported through rate increases. There are some studies that indicate that the public is willing to pay for reliable, high-quality water services (Howe & Smith, 1993, 1994; AWWARF, 1998). However, water managers and city councils and officials often lack the political will to increase prices or to practice cost-based rate-setting. Political leaders fear the consequences of unpopular policies, and there are legitimate concerns regarding the affordability of an essential service to poorer segments of the population. Although economic efficiency is promoted when water service rates reflect the true costs of providing those services, the prospect of higher rates is often substantial barrier to privatization.

Water System Regulation

All community water systems in the US are subject to regulation by state drinking water agencies pursuant to the federal Safe Drinking Water Act (SDWA). Systems must meet federal standards, but states can impose additional standards. US states have primacy with respect to water quality regulation, including regulation of withdrawals and diversions.

The rates and profits of investor-owned utilities are economically regulated. This regulation by states is regarded as a substitute for competitive markets and public ownership, which presumably ensure competition by other means. Given their physical and economic properties, competition between water utilities is highly limited and they thus tend to be monopolistic. For publicly owned monopolies, accountability is ensured through electoral and related public channels. For privately owned monopolies, accountability is ensured through economic regulation by US state regulatory commissions. Economic regulation applies to virtually all private water utilities (but not private contractors), and some publicly owned systems. State public utility commissions regulate prices charged by public utilities. Jurisdiction of commissions varies between states.

Economic regulation may account for some of the differences in performance between public and privately owned water utilities. Regulatory methods that can encourage privatization include modified rate-making and profit-related incentives such as profit sharing. To date, there has been limited state-level regulatory review of water services privatization agreements in the US. To many privatization advocates, regulation is not necessary because local governments can regulate through the contract. Others feel that regulation is necessary to prevent abuses of monopoly power and to ensure that cost savings are passed on to customers.

Conclusions

The present and future needs of the water treatment and distribution systems in the US are tremendous. The resources necessary to maintain, repair and upgrade drinking water and wastewater treatment facilities are not always readily available from the public purse, and public officials are often reluctant to accept the political consequences of raising taxes or fees to help cover these costs.

Privatization of water services represents a viable alternative in many instances. This privatization takes many forms, ranging from the contracting of services such as meter reading and laboratory analysis, to the transfer of assets to the private sector. No one form of privatization best fits all situations, and privatization agreements and contracts should be tailored to a water utility's and community's unique circumstances. Moreover, privatization should not be equated with competition; competition exists primarily during the contract bidding process. Once the contract is awarded, the contractor is not fully subjected to the forces of the market, but rather the conditions specified in the contract.

Privatization, however, does not represent a panacea for addressing all water utility problems, and not all water privatization efforts in the US have been successful. Privatization efforts have in some instances led to repossession of contracts and litigation, and some contracts have not been renewed. By the same token, some public water utilities have experienced similar problems. Well-run and poorly-run organizations exist in both the public and private water sectors.

Small to medium-sized water utilities face the greatest challenges, often have limited resources and access to contemporary facilities and training opportunities, and therefore are prime candidates for some degree of water utility privatization. Such communities could also take advantage of services from the public sector. Regionalization, or consolidation of multiple smaller water utilities, often represents an attractive option, and can be accomplished through both the private and public sectors. But there are barriers to privatization, including possible loss of control of a vital public service, possible loss of jobs, and the resources required to manage the contract bidding process. Barriers to entry may also exist for contractors, as contractors need to be assured that the bidding process will actually result in a contract being signed. Contractors also do not possess the ability of municipalities to issue tax-free bonds.

An important effect of the prospects for privatization in the US has been to motivate public water utilities to improve their performance. Several competent multinational firms are looking to increase their share in the US water market, and many public utilities have responded to this challenge by instituting practices such as 'benchmarking' and 're-engineering' to help raise performance levels.

Despite the presence of sophisticated and experienced multinational and US firms in the water utility sector, continued public ownership and operation is the most likely future for the majority of water utilities. But despite an increased interest in the prospects for water services privatization, the approximately 14% market share of private firms in the US water business has remained steady over the past 50 years and is likely to do in the near future.

References

AWWA (2001) *Reinvesting in Drinking Water Infrastructure: An Analysis of Twenty Utilities' Needs for Repair and Replacement of Drinking Water Infrastructure* (Denver, CO, AWWA). Available at http://www.awwa.org/govtaff/infrastructure.pdf (accessed 11 March 2002).
AWWARF (American Water Works Association Research Foundation) (1998) *Consumer Attitude Survey on Water Quality Issues* (Denver, CO: AWWA Research Foundation).
Beecher, J. A. (2000) *NAWC Forum Paper* (Washington DC: National Association of Water Companies).

EPA (US Environmental Protection Agency) (1990) *Public Private Partnerships for Environmental Facilities: A Self-Help Guide for Local Governments* (Washington DC: US Environmental Protection Agency).

EPA (1997) *Drinking Water Infrastructure Needs Report—First Report to Congress*, EPA 812-R-97-001 (Washington DC: US Environmental Protection Agency).

EPA (1999) *EPA Safe Drinking Water Information System Factoids: FY 1999 Inventory Data*. Available at http://www.epa.gov/safewater/data/99factoids.pdf (accessed 11 March 2002).

Howe, C. W. & Smith, M. G. (1993) Incorporating public references in planning urban water supply reliability, *Water Resources Research*, 29(10), pp. 3363–3369.

Howe, C. W. & Smith, M. G. (1994) The value of water supply reliability in urban water systems, *Journal of Environmental Economics and Management*, 26, pp. 19–30.

National Research Council (2002) *Privatization of Water Services in the United States: An Assessment of Issues and Experience* (Washington DC: National Academy Press).

PWF (Public Works Financing) (2001) Water privatization scorecard: communities with long-term water partnerships, *Public Works Financing*, 147 (January), p. 8.

Scalar, E. D. (2000) *You Don't Always Get What You Pay For: The Economics of Privatization* (New York: Cornell University Press).

The Water Industry Council (1999) Survey report: public attitudes about tap water, wastewater treatment, and privatization (draft).

US Conference of Mayors Urban Water Council (1997) *A Status Report on Public/Private Partnerships in Municipal Water and Wastewater Systems: A 261-City Survey* (Washington DC: Conference of Mayors' Urban Water Council).

WIN (Water Infrastructure Network) (2000) *Clean and Safe Water for the 21st Century* (Washington DC: Water Infrastructure Network).

Price Setting for Water Use Charges in Brazil

RAYMUNDO GARRIDO

Universidad Federal da Bahia, Faculdade de Ciencias Economicas, Salvador, BA, Brazil

Introduction

Both the price setting for use of bulk water sources as well as the study of the outflows of these water sources depend on theoretical-scientific knowledge, which sets levels, based on its own regulations and methods, at a reasonably precise rate for prices to be used in these peculiar markets. For this reason, and also in anticipation that a broad range of charges for water use will be implemented across Brazil, the paper intends to discuss some of the elements involved, and looks at studies offered by different researchers in this area, in order to contribute to the planning sector and water resource management.

There is also a debate on the capacity, or lack of it, for market mechanisms, without interference or the imposition of rules, that will better serve society's needs, either collectively or individually. This evaluation is extremely useful to show the 'rationale' of public intervention in allocating water resources to competitive stakeholders.

General Classification of Charging Methodologies

The greatest difficulty in attributing prices for water use charges lies in its various uses, requiring the application of criteria for price differentiation.[1]

There are several price-setting methodologies for a public good such as water. It has the characteristic of its mobility as well as being used for distinct uses. For this reason, these methodologies are almost specifically for water resources, not being useful for other natural resources, unless they suffer significant adjustments. Carrera-Fernandez (2000a) in his excellent research, classifies these methodologies into three groups: (1) optimization

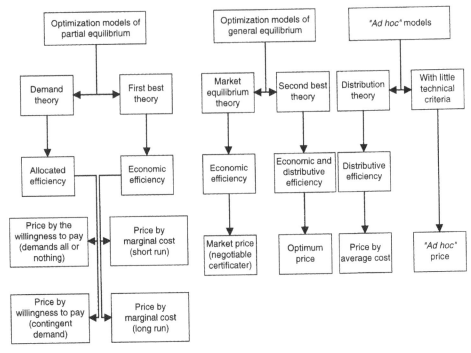

Figure 1. Methodology classification for calculating price to be charged for water use. *Source:* Carrera-Fernandez (2000).

models with partial equilibrium; (2) optimization models with general equilibrium; and (3) ad hoc models, according to Figure 1.

It is important to note that the methodologies for the first two groups are price-setting mechanisms. These methodologies adopt the optimizing behaviour of the economic agent able to take rational decisions, encouraging the price for water to be according to hypotheses which are widely accepted by economic theories.

Concerning the third group, the ad hoc models, those that do not have a legitimate process of economic optimization, the price to be charged corresponds, in most cases, to the average cost of production, simplifying the calculation. In spite of encouraging distributive efficiency, charging based on average cost may distance itself significantly from the socially optimum value, that is to say of the marginal long-term social cost.

The following sections briefly comment on some of the methodologies proposed in Brazil, suggesting as an adequate method, the 'Second Best' theory, for its capacity to optimize the price level and at the same time induce the system to an efficient allocation of water resources, as well as to internalize social costs, reflecting the true cost of water for each use and promote the creation of a fund for programmed investment in the watershed.

Brief Analysis of Some Models Proposed in Brazil

There are innumerable studies on charges for water resources use in Brazil, but only in the state of Ceará has one of these studies been put into practice.

Research for the São Paulo State Rivers

The study developed by FUNDAP for the São Paulo state government is worth mentioning. It took as its reference the basins of the rivers Piracicaba, Capivari and Jundiaí. A great virtue of this research, from an economic point of view, is the fact of recognizing explicitly that the water resource sector—management and operation—is a renewable resource industry, although limited, whose complexity of operation has increased due to the competition among the multiple uses of water resources.

The methodology used in this research is based on the user-payer and polluter-payer principle, by dividing investment costs between the various users, that is, adopting the price to be charged equal to the average cost. Nevertheless, for the dilution of effluents there is a clear preference towards price determination based on long-term marginal costs, not including the management costs of its own sector in the calculation. Equal treatment is given to industries, in the case of abstraction water, by adopting equal long-term marginal costs, based on an estimate of investments to regulate the flow.

Another practical aspect is a look at the possibility of the state, or Federal Union, initially subsidizing part of the investment, with a programmed withdrawal of this subsidy after a specific period of time when the administration of the basin would be able to be self-sustainable.

The Waters and Electric Energy Department (DAEE) pursued this research and signed an agreement with the National Engineering Consultants Consortium and the Economic Research Institute Foundation (FIPE) to study the implementation of charges in the state, especially in the Piracicaba, Alto Tietê and Baixada Santista regions. The FIPE/CNEC studies divided water usage into two parts for charging purposes: the abstraction and the consumption of water. For these and for hydroelectric power, the methodology used was willingness-to-pay[2]. In the case of water use for effluent discharge, the average cost to maintain the quality of the water bodies in the desired standards was used.

The results of these studies are illustrated in Table 1, including not only the price to be charged for each water usage, but also the influence of the charge on the revenue of the basins studied. The study shows both prices for the critical basins chosen and it determines prices for the whole of the state of São Paulo, according to the last line in the Table.

De Souza (1995) also undertook research in São Paulo, examining the basins of the Jaguari, Atibaia and Piracicaba Rivers. The research looked at the discharge of effluents based on the average cost of treatment for these effluents through anaerobic lakes, optional lakes and systems that remove the nitrates. The study was based on statistical adjustment

Table 1. Prices for the use of water and potential revenue in river basins in São Paulo state

Basins	Impounding (R$/1000m^3)	Consumption (R$/1.000m^3)	Electric energy (R$/mWh)	Pollution (R$/tonDBO)	Revenue (R$ 1.000)
Alto Tietê	7.00	31.00	4.00	320.00	145.561
Piracicaba, Capivari and Jundiaí	3.00	17.00	4.00	193.00	31.111
Baixada Santista	1.00	6.00	0.10	228.00	9.421
State of São Paulo	3.00	13.00	0.10	278.00	–

Source: FIPE/CNEC (1995a).

for curves generated by the charge and separated the discharge of effluents into water bodies from the consumption of water from water bodies. It is interesting to note that for the first part of the problem that was studied, de Souza (1995) considered as cost factors issues such as impoundment, effect of dilution, and dilution capacity of the surface water body. This means that part of the quality control of the water body may be carried out by the pricing policy for consumption use of the same water source.

Table 2 presents the results of this research, in terms of average price per US$/(habitant year), achieved through simulations for the São Paulo Metropolitan Region and for the city of Piracicaba. Analysing the values achieved for the city of Piracicaba, which were divided into consumption and discharge, this method underestimates the value of water for consumption in order to benefit the domestic user by charging reasonable prices, and will always pay a value that has little significance no matter what the level of saturation.

'Saturation state' is the depuration capacity of the water source, which means that prices will increase where the state is at a critical level. There is no doubt that this research carries relevant technical criteria while taking into account the depuration capacity of the water source. However, the research may be broadened to include user and societies' (demand function) requests, aggregating variables that allow an optimizing criteria behaviour, be it to minimize costs, or to maximize social well being, such as that discussed in economic literature.

Research for the Rio Grande do Sul State Rivers

The research elaborated by Lanna (1995a,b) for several stretches of the basin of the Vacacaí River, adopted the charging method of long-term marginal costs for water use in rice agriculture by flood irrigation. Investment, operation, system maintenance, enterprises and necessary procedures were all taken into consideration to determine the cost of water supply.

The analysis of the results obtained in the Vacacaí River are illustrated in Table 3, noting that the marginal cost of bulk water, known here as a function of supply which aggregates the elements mentioned above, is more sensitive to a variation of the discount fee than to a variation in time investments. This means that the marginal cost of water use has a greater reduction if the discount fee is reduced from 11%[3] to 8%, than if the time factor of investment amortization increases from 20 to 50 years.

In another study, Lanna & Pereira (1996) developed a charging methodology for the Sinos River. Unlike the case of the Vacacaí River, it seeks to minimize the charging

Table 2. Average prices for domestic water use in São Paulo (US$/habitant × month)

	Saturation state	
	Critical	Normal
Human Supply, Metropolitan Region of Sao Paulo	29.00	0.11–0.25
Consumption, Piracaba	0.62	0.01
Discharge, Piracaba	41.50	1.79

Source: de Souza (1995).

Table 3. Long-term marginal cost for the Vacacaí river basin

Dam	Investment US$1000	Regul. outflow 1000m³/year	Amortization in 20 years		Amortization in 50 years	
			Annuity US$1000/year	CMg^{LP} US$/1000 m³	Annuity US$1000/year	CMg^{LP} US$/1000 m³
VAC-4	8.416	48.599	1.057 (857)	22 (18)	931 (688)	19 (14)
VAC-6/7	7.982	43.865	1.003 (813)	23 (18)	883 (652)	20 (15)
Pondering Average (US$/1.000m³)			–	22.5 (19.0)	–	19.5 (14.5)

Source: Lanna (1995a,b).

distribution costs in relation to pollution control costs. The minimum basic prices are determined in order to maximize a minimal relation between the basic price of each parameter (or pollutant) and the excess of emission for a given standard, subject to an inverse relation between the control costs of each source of pollutant and its charging cost. The basic minimum prices vary positively with the cost and necessity to control the pollutant. This means that the difference between basic minimum price and the excess of emission, although minimum, is arbitrary. Consequently, the volume of required resource to finance necessary investments to control the pollutant source is also arbitrary, since the volume of resources determines this difference. The result of the application of the method is the impact of the charge over operational cost for the water user, as shown in Table 4.

Although this method contributes to a fair charge cost distribution among users, nothing ensures that it will be efficient in allocating water resources, nor the minimization of social costs. In this manner, such prices seem to contribute to the creation or increase in distortions while allocating water resources among interested users.

Research for the Paraíba do Sul and Doce River Basins

The research for water resource use charges in the Paraíba do Sul and Doce river basins, developed by FIPE in 1997, proposed two main goals: (1) to establish the real value of this natural resource for the user; and (2) to promote the rationalized use among multiple users. The methodology was based on average cost, by the dividing or sharing of the investment costs. The text of the available volume only has prices for the Paraíba do Sul.

Even though the methodology presents practical advantages, it is not possible to confirm that it is able to assure an efficient allocation. In other words, it does not guarantee the best technical and alternative uses for water resources, since the determined prices distance themselves from marginal costs. Analysing different production and consumption models, as well as different market structures, the study also evaluates the charge application impacts over segments of water users, in an exclusively partial equilibrium.

Table 4. Impact of the charge for water use due to pollution by the industrial sector in the Sinos river basin (% of operational cost)

Sectors	Scenarios		
	1	2	3
Leather, skins and similarities	0.20	0.20	0.21
Drinks and alcohol	0.02	0.02	0.02
Textile	1.61	1.63	1.66
Foods	1.40	1.42	1.45
Chemicals and petrochemicals	0.00	0.00	0.00
Iron and steel metallurgy	0.0002	0.0002	0.0002
Cellulose, paper and pasteboard	0.0003	0.0003	0.0003
Public unit	1.40	1.42	1.45

Source: Lanna & Pereira (1996).

The research is based on the user-payer and polluter-payer principle and the expenses from the investment plan were divided among multiple users to balance future costs and revenues. Although this is a fair argument when considering its distribution, there is a problem in the way these costs are passed on to those who benefit from this system, for it is possible to strongly penalize some users with less means to cover such costs, while others who are better off and more able to pay, benefit by paying a much smaller amount.

The determination of demand behaviour was based on a questionnaire of the users' willingness-to-pay for an additional utility unit. The supply function, however, was taken as a vertical curve at the level of the water availability of the system. This means that, since the supply curve adopted is inelastic, the research does not recognize that water resource management is in fact an industry, in the whole sense of the word, which produces goods and/or services for its users. In micro-economic terms, the supply curve is no longer represented by the social marginal cost as is usual in balance models where production—in this case, the production of bulk water—varies according to a greater or lesser effort of production, effort which is reflected in the levels of varied costs. In this manner, the method truly used by FIPE in its research was the willingness-to-pay, adjusted to finance the values of the investment plan. The prices-to-pay, calculated by FIPE, are shown in the first column of values in Table 5, taken from the research carried out by the ICF Kaiser-Logos Consortium, which compares these with the prices from the state of São Paulo (second column of the same Table), and with prices from the Adour-Garonne basin (France). The prices for the Adour-Garonne basin are not reproduced (into what would be the third column) because this basin does not mirror the conditions of the Paraíba do Sul basin.

FIPE's research also studies the value for water use as a means for dilution and effluent distancing, adopting a different procedure than was used for the consumption use of water resources. In the case of effluent discharge for dilution, the study considered the opportunity cost of controlling emissions and rejections as an alternative to costs incurred during the water treatment process of the water sources after they received effluents.

In Table 5 the discharge effluents are separated into oxidizing materials and solids in suspension. The calculation process, based on opportunity cost of discharge control, compared this cost with what could be internalized by the pollutant agent (water user) and the costs incurred by the water treatment process of the water source if it discharges without treatment. If this happened, once the water source was polluted, the user would demonstrate a reduction in willingness-to-pay, causing a lowering of the demand curve. In this case, the loss of economic value due to the contamination of water would be given by multiplying the quantity of water by the difference in price, considering a supply curve (inelastic, as observed), but with a minimum value lower than there would be supply.

Table 5. Price comparison for water use

Generating fact	Paraíba do Sul	State of São Paulo
Abstraction (US$/m³)	3.60	3.10
Consumption (US$/m³)	41.90	12.70
Oxidizing materials (US$/ton)	147.80	278.00
Solids in suspension (US$/ton)	124.50	–

Source: ICF Kaiser-Logos Consortium (1998).

Research for the Ceará State Rivers

Lanna (1994) and de Araujo (1996) developed studies for bulk water use charges in the basins of Ceará, both based on average costs. However, for comparison they used a methodology developed by the National Irrigation Policy during the analysis, based on the long-term marginal cost.

These studies introduced two innovations. The first was the exchange of the total volume of water effectively used by agricultural users with the regulated volume in the basin. The second was the definition units of the area to be irrigated by each user, taking into consideration the relation between water availability and the irrigated areas. The goal of these changes was to distribute the costs in an equitable way, therefore penalizing the large landowners who used little water, and benefiting small landowners who irrigated much more. In this case, the policy to be adopted by charging would be like a tax on large landowners, which could have a negative effect on agricultural production in the state. The distribution distortions that need to be corrected would be better off, probably, in this case with the agrarian instrumental policy.

In the calculation process the necessary investment was made to ensure regulated outflow, leaving the management costs to be covered by the state government, with the argument that evaporated water as well as that not used by the system should not be paid for by the user. The research also introduced a cross-subsidy, calculated by irrigated area, favouring the small landowner farmer instead of a large landowner, taking into account the criteria of considering the irrigated unit of each owner in proportion to the water availability of the basin.

It appears that in order to adopt the charging tool for water use to favour the small owner, even as a measure of imminent social character, the state of Ceará may, inadvertently, be promoting distortions in the allocation of water among various irrigators by encouraging extravagant consumption, since it will be stimulating rural producers to use all irrigable land, even if the land is not appropriate for such use.

As well as the mentioned studies, the state of Ceará has greatly advanced in the last three years by applying a charge for water resource use. The Water Resource Management Company of the State (COGERH) started the process of charging in 1996, beginning with the Metropolitan Region of Fortaleza, for industry and human supply. Table 6 shows the prices being charged today in Ceará for urban supply, industrial supply in the whole of the state and irrigation from the Trabalhador Channel and the Acarape River.

In 1999, a total of approximately R$6 million was collected, with 50% distributed for supply and 50% for industry. Relating that to present levels, human supply outflow reaches 7.42 m^3/sec and 0.14 m^3/sec for industry, which seems to correspond effectively only to

Table 6. Prices charged for water use in Ceará

Use	Price (R$/m^3)
Urban supply	0.013
Industrial supply	0.67
Irrigation from the canals	0.02
Irrigation from the Acarape River	0.004

Source: COGERH (2000).

the charge in the Metropolitan Region of Fortaleza. With regard to the dilution of effluents, charging has not begun either to CAGECE or to industries. However, at present industries send their effluents to the Sanitation Company which treats the discharge and returns it to the river bodies. Even though CAGECE does not pay COGERH for any of the effluent discharge, it charges R$1.92 per cubic metre of received effluent.

Research for the Bahia State Rivers

The studies used for water resource charging in Bahia began with the methodology developed from 1991 for the Fêmeas River basin, a second order tributary on the left margin of the São Francisco, where there were serious conflicts between irrigators and small electric energy generators.

Later, several aspects of this methodology were broadened, corrected and improved by Carrera-Fernandez (1996), with studies of the Itapicuru and Paraguaçu river basins. All the prices in the state were proportioned according to the optimization process, briefly described later in the paper for the Vaza-Barris river basin.

In the price optimization process, the long-term marginal cost was adopted to determine supply, which included all the estimated investments. Such investments were defined with the approval of the master plan of each basin or hydrographic region, by the Water Resource Superintendency of the state government. The study also showed what would be gained by collection, per basin, when the charge was implemented. In addition, a calculation was made of the consequences of prices for final products of water use activities in each basin, in other words, how the charge for water resource use would affect the commercial and economic activity of each user-payer category. The data on how the prices of the final products were calculated for each water user are not included in this paper due to lack of space.

Charging in a Rationing Regimen

The charging methodology based on the willingness-to-pay while in a rationing regime has been backed up by Kelman (1997), and is based on a process similar to an auction of goods. The model claims that any number of users in a set will take their decision of water use to minimize the expense function $g[a(i)*]$. The term $a(i)*$ represents the water utilization coefficient given as a concession to the user 'i', which comprises two parts: the first corresponds to the value of the bulk water bill of any user, whose value is unknown; and the second, is relative to the loss caused by an eventual water rationing, which is only felt by the user himself. In practice, each user declares his rationing cost, which indicates his willingness to pay.

As a priority rule in rationing it is interesting to notice that, given there are safeguards for users of 'insignificant outflows'[4], this method proposes that the criteria for distribution prioritizes the user who proposed the highest price before the rationing. In other words, the user who proposes the highest value to pay during the rationing will be the last to feel the rationing, and vice-versa. It is as if there were a line in which the priority for use would be given to the users who declared the first higher costs of rationing, and apportioned the requested volumes, until the capacity limit of the water source under the rationing circumstance.

The method reflects the result that should appear in the marginal price setting analysis when in a rationing regime in a river basin, because the supply prices to be paid by users, even in the limiting condition of rationing, are an essential element for determining the demand[5] function. Therefore, the method is extremely useful to confirm, or eventually cancel, the studies that should be developed by the basin agency to serve as a subsidy to the committee. The committee will promote the rounds of negotiation among the ones that are interested. Strictly speaking, negotiation is used here to indicate the liberty that should be present in a committee's session to decide on the supply prices of the auction, which is the ritual recommended in this type of method.

Another interesting aspect of the proposal is that the value collected should be applied to reimburse the users who suffered rationing because the price they offered was not sufficient to guarantee their part. It is interesting to observe that the price to be paid by each user corresponds to the price offered by the last attended user in line. In other words, the user who offered the greatest value, or rather, the first in line, will pay a price lower than his own bid. This will also happen with all of those who are in line until the next to the last one. The others will suffer the rationing and the reward of being reimbursed with the product of the collecting.

Some authors say the payment made by each user should be in accordance with his bid, so that the collection from each basin would increase, and more important, reduce speculation, since the speculative users could increase their offer only to win the bid and then later pay the least. It is important to notice that the user who bids a higher value simply to be the first in line will eventually also have to pay a higher price than the rationing cost, since he himself will have contributed to increase the price level that will be paid by all users. In addition, if the process figures on a payment from each user according to his bid, another type of speculation may arouse: that the first users of the non-rationed part of the line, which made the highest bids, may buy the rights of the last ones in line (that made low bids), creating a non-healthy price system.

What must happen in the practical application of this method, is that the amount collected in a single price should be able to reimburse the users that have suffered rationing, as well as to suffice the basin demands in terms of covering expenses and making investments, which are important, and overall in rationing periods. This is not absolutely guaranteed in terms of Kelman's proposal. For this reason, the basin agency will announce, before the auction, a minimum price that will be charged. There is no doubt that this price level corresponds to the long-term marginal cost taken at the level of available volume at the time of rationing. This consequently reinforces the meaning of the price setting methodology, which is accepted according to the optimizing behaviour of the water user while an economic agent.

The proposed methodology meets up with innumerable qualifications in order to receive the merit of being recommended to be put into practice, for it will serve, at the same time, for the fair decision making over outflow apportioning under rationing conditions, and to support the previous studies undertaken by the basin agency, or even to indicate the necessity to change these studies.[6]

Finally, the method does not reflect the behaviour of the users in normal conditions, that is, in non-limiting conditions such as rationing. However, it is essential to recognize that while estimating marginal costs for the price setting models, a decision should be made on marginal rationing cost, as indicated by the proposed method since it expresses more fairly the social cost of water.

Case Study of the Vaza-Barris River Basin (Bahia and Sergipes States)

The study on water use charge for the Vaza-Barris basin, whose main stream flows through the states of Bahia and Sergipe, was initiated by the Secretariat of Water Resources of the Ministry of the Environment, due to the severe drought in 1998 and 1999. The level of the Cocorobó dam lowered to an almost empty state in the state of Bahia. The study follows the methodology developed by Carrera-Fernandez (1999b), of price optimization based on the Second Best theory, which seems to best fit the characteristics of the Brazilian economy, of presuming distributive efficiency as a first order condition, charging the fair value for each water user.

The starting point was the water demand study in its various uses. The ordinary demand functions of the various user segments, as well as the corresponding demand price-elasticity were obtained by 'all-or-nothing' demand, whose theoretical base is presented in the next section. The marginal cost was established according to the marginal unit cost for the short and long term.

The study produces three types of prices: the demand price, which is the price given by the ordinary demand function; the optimum price, which is divided into optimum prices with or without investment restrictions; and finally the reserve price, which represents the maximum value that users are willing to pay for water use from a given water source and remain indifferent as to impound water or search for an alternative solution which produces the same effect. This reserve price is given by the 'all or nothing' demand curve. The results of the study are illustrated on Table 7.

Indications so far suggest that this seems to be the best methodology for establishing prices to be charged for water use since it is the only one able to fulfill all requirements mentioned previously. For this reason, further comments are focused on the study of price optimization for water resource use, due to the suggestion that this method be considered a task for the future basin agencies, who will subsidize their respective committees when negotiating between interested parties for the establishment of price levels to be charged (for detailed analysis of several methodologies, see Garrido, 2003).

Table 7. Prices paid for water use in the Vaza-Barris river basin (R$/m^3 and R$/kgDBO)

Uses	Demand price	Optimum price		Reserve price
		Without restriction	With restriction	
Human supply[a]	0.247	6.14×10^{-2}	1.90×10^{-1}	0.49
Industrial supply[a]	1.300	1.75×10^{-1}	2.32×10^{-1}	2.74
Irrigation[a]	0.005	$1.73 \times 10^{-1*}$	9.54×10^{-3}	9.54×10^{-3}
Dilution for sanitary discharge[b]	0.020	1.05×10^{-2}	3.72×10^{-2}	0.04
Dilution for industry effluents[b]	0.205	2.51×10^{-2}	4.58×10^{-2}	0.41

Source: Carrera-Fernandez (1997b).
Notes: [a]R$/m^3; [b]R$/kgDBO; *Value greater than the reserve price.

Table 8. Optimum prices for water use in the Pirapama river basin (in R$/m^3 or R$/kgDBO)

| | | With restriction | |
Uses	Non-restricted	Payment capacity	Without electric power generation
Human supply[a]	1.68×10^{-2}	2.13×10^{-2}	2.15×10^{-2}
Industrial supply[a]	8.85×10^{-2}	1.04×10^{-1}	1.04×10^{-1}
Irrigation[a]	8.86×10^{-2}	9.51×10^{-3}	9.51×10^{-3}
Electric power generation[a]	3.36×10^{-2}	2.24×10^{-3}	–
Fertirrigation[b]	1.31×10^{-2}	1.57×10^{-2}	1.58×10^{-2}
Dilution of industrial effluents [b]	8.88×10^{-3}	1.09×10^{-2}	1.10×10^{-2}
Dilution of sanitary discharge [b]	2.62×10^{-3}	3.34×10^{-3}	3.38×10^{-3}

Source: Carrera-Fernandez (1999a).
Note: [a]R$/m^3; [b]R$/kgDBO.

Study for a State of Pernambuco Basin: The Case of the Pirapama River

The Pirapama River basin has been used as a source of water supply for the city of Recife since 1918, when the Gurjaú dam and channel for Recife was built. Studies carried out by COMPESA in the 1960s and later in the 1980s revealed that the Pirapama River was the main water source with the potential to serve the demands of the metropolis. Despite the priority for public supply of the Pirapama, the waters of this basin have also been used for other purposes, e.g. supply to the rural zone by means of wells, industrial supply, hydroagricultural use and hydroelectric power.

The water use charge studies of the Pirapama River were developed by Carrera-Fernandez, according to the price optimization method, based on the Second Best theory, which has previously been explained. The results obtained are presented in Table 8.

These studies have come at an extremely convenient time because the Pirapama dam reconstruction is due to be finished, after several years of interruption. The deficiency of this water source leaves a great void for the Recife Metropolitan Region (RMR), which lives with severe rationing due to insufficient water sources. Today, RMR relies on the Tapacurá, Goitá and Carpina dams, which were originally built to retain floodwater, and not to be used as water sources. There are also other water sources that supply the city of Recife and its surroundings, fulfilling no more than 60% of the requirements of this metropolitan region.

Study of Reservoir Water Tarification: França Dam/Jacuípe River, State of Bahia

Studies have been developed to determine water resource use prices for the França reservoir (Bahia), with the aim of indicating a guarantee of the financial sustainability for use management of the reservoir as well as recuperating its investment.

The research methodology was based on the applied statistics of hydrology, which recognize the fact that water stored in a reservoir may have two distinct behaviour values, which together with the operational and maintenance costs of the reservoir (CO&M) will form a final price (gross price) to be paid by the user. These two parts are: the water use value (V_u) and the preventive value (V_p). The preventive value corresponds to costs included in maintaining a determined volume of stored water, in order to prevent users from suffering a lack of water. This would be equivalent to an insurance against the lack of water.

This method tries to determine the value of water in relation to its stored volume. For V_u, it is admitted that for each unit of volume taken in a month in which the reservoir is full, it will cost less than when the reservoir is almost empty. The behaviour of V_p is inversely proportional to V_u, in other words, when the reservoir is full the value will be high and vice-versa.

The total value to be paid per month corresponds to:

$$P_{gross}(R\$/1000\,m^3) = (V_u + V_p + CO\&M)/I_i,$$

where I_i is the total effective withdrawal of water from the reservoir in a given amount of time i (month). The methodology advances with the value setting of V_u and of V_p, establishing, by algebra transformations of the concepts of these two realities that: (1) $V_u = \alpha.I^2/Q_i$, where α is the consumption coefficient (modelling parameter), given in monetary units per volume and Q_i is the water being stored in a month; and (2) $V_p = \beta.Q_i.(H_i/H_{max})^{(\theta-1)}$, where β is the storing coefficient (modelling parameter), H_i is the height used in the dam for storing in the period i, H_{max} is the heigth of the dam, and θ is the parameter that varies with the type, form of dam, inundated area and specifics of the enterprise.

Once the concepts above are established, the method consists of applying a *Cascata* program, developed by Professor Eduardo Lanna (UFRGS/IPH), which allows the determination of the function which relates storing capacity of a dam with annual discharge guarantee for this same dam. For this, a few boundary limits are established, such as, for example, the percentage of V_u, that should be channelled to the basin agency, as well as a range of price levels 95% of the time. With these elements prices are reached that are a function of the stored volume. In addition, the method also extracts internal private return for distinct periods of time.

This is an interesting exercise, indicating management measures to adopt according to the behaviour level of the reservoir. It appears the method could advance even more, once the functions are established for V_u and V_p which are actually functions of production, and not of cost. The transformation from a production function to a cost function was made by the authors through parameter arbitration, a and b, whose units introduce, by means of dimensional analysis, the monetary factor directly in the production function. It would seem more reasonable if the cost function were obtained by determining the optimization of the production function system subject to budget restrictions, operating Lagrange's undetermined multiplier. The true total cost function would be obtained, which would originate in the marginal cost, as a starting point for the adequate analysis of price setting.

The comments offered here do not, however, take the merit of the methodology developed, since it correlates the behaviour of innumerable technical parameters of reservoir operation. On the other hand, it serves as an incentive for its own improvement by introducing the necessary elements in Microeconomic Analysis.

Conclusions

The price formation study is one of the most interesting chapters, and also the most complex, of the Microeconomic Analysis. When the matter of transaction is bulk water use, the complexity of the question increases. This is a natural resource which is naturally available that assumes the form of its recipient body, flows in physical space due to gravity, and may be stored either on the surface or underground, and it brings peculiar ingredients to the process of calculation, which make the establishment of the water supply function a hard task. On the other hand, the water demand study is no less complicated. First, in many cases there are conflicts between economic agents that use these resources, and the formulation of budgets that consider the availability and price of each method of production, water being one of them. Second, but no less important, is the behaviour of the final consumer of the resources, which plays an important role in decision making. Note that the consumer, in general, and depending on the relative position of the economic agent in the intersectorial relation, is the familiar unit, in other words, is the human being, which is a social being. This is where the petitioning and consumer theory laws enter, which not only have their tastes and preferences, but also their income and prices of similar products, as well as the three indicating elements of search behaviour.

Continuing the series of elements that enter the charging subject is the concept of price and social nuance (social prices, economic prices and shadow-prices), which include externalities created by production and consumption decisions. This aspect is particularly important in the water resource management sector, since the decision to use one more cubic metre of water by the user imposes to others externalities that need to be included in the price to be paid. Along with this are the aspects already mentioned. The price to be charged for bulk water use is also seasonal, increasing in dry seasons and decreasing in rainy seasons. It may also vary in terms of location, being, in general, lower in downstream areas and higher in upstream places of the watershed.

Price setting for different water uses is also an important consideration when formulating public policies for water resources management. Together these public institutions involve not only formulation agencies, but also implementation entities of these policies, as well as interaction with a series of non-governmental entities which harmonize the institutional outline of the water resource segment, or rather what has brought the most innovations to public management in Brazil. In this are included the basin committees, the basin agencies, the civil water resource organizations, as well as state and federal management agencies, as well as the National Water Agency. All these elements keep a strict relation with the water resource policy tools, especially with the water use charge.

An interesting aspect, not mentioned is that of the physical space defined for the charging application. Brazil has several basins that are several million square kilometres in size. Others, even though they are smaller, are still larger than many countries of the world. Today, in Brazil, there is a tendency to not be able to manage water use in such large areas. In São Paulo, continuous discussions have admitted that the most comfortable mode for this management would be one that would allow the manager or committee member to move from his home to the most extreme point of the watershed, work during the day and return home at night. Among this empirism, it has been discussed that in developed regions, where there are paved roads and communication facilities, a square whose side is equal to 200 kilometres could be an ideal module for water resource use management. Such dimensions would equal $40\,000\,\mathrm{km}^2$, which would be the maximum size for the basin and sub-basin division aiming towards its management.

Based on this criteria, the Brazilian territory could be divided into sections, maintaining the principle of water divisors, to form watersheds whose main rivers could be third, fourth or even fifth in order in relation to the main basin, to have better spacing of management and application of its tools.

It is still probable that charging may resist the module suggested above. It is possible that the areas where water tariffs need to be established should be considered very specifically from the other areas within the same basin. For this to be possible, it is necessary only that the river be long enough, to pass through various regions with distinct characteristics and water uses.

Another issue to be considered is the possibility of awarding water rights that can be transferred among users, as long as they have the appropriate concessions. Although this is mentioned in Figure 1 as one of the price setting systems, the theme has not been explored in this paper since it could be developed further in a future study. Such a market favours artificial reservoirs, and in these circumstances the externalities caused by the upstream users over those downstream practically disappear. It is interesting to note that when the reservoir is built for the generation of hydroelectricity, the electricity, in this case, is a consumption use in relation to the other multiple users of the reservoir.

Finally, it is important to stress that the price setting study for bulk water use comprises an interesting tool for economic planning and public policies. The simple analysis of a broad matrix of input-output may reveal that consumers (or users of produced goods) of goods such as drinks, leather goods, paper and cellulose goods, as well as a large number of other products, are indirect consumers of water sources. Therefore, the water resource management policy of critical watersheds should take into account the impacts caused by sectors that, indirectly, use bulk water. This is also related to charging for use of this natural resource. The aim of this charging methodology analysis was to emphasize the importance of price setting for water use, an essential basis for debate among basin committees, where negotiation is a management tool that effectively prevails over all others, and will proceed with the prices of transactions among water resource use.

Notes

1. In economics, price differentiation implies using the concept of price-elasticity demand, and confirms, with precision and via prices, the unequal treatment of stakeholders with unequal economic situations.
2. The demand curve was obtained by the contingent evaluation method, by means of combining open and closed type of questions. The research considers, although imprecisely, the willingness-to-pay as a surplus measure of the consumer. The surplus of the consumer is defined by the difference between the maximum value and that which the consumers are willing to pay and the value that they effectively pay for a greater quantity of the good or service.
3. The author adopted the rate of 11% because of the opportunity cost capital.
4. This concept will vary from basin to basin in terms of the outflow level which includes the demand for human supply and that to small users in rural areas who use water for their own sustainability.
5. The accepted definition for price offers by users in rationing situations is done through a study of demand 'all or nothing', applied to the methodologies *First Best* and *Second Best*.
6. Note that price definition goes through the basin committee and, in the case of this methodology, performs the auction. However, the possibility of the basin agency developing price setting studies should not be disregarded as a starting point for the negotiation to be developed in the committee's environment.

References and Bibliography

Altaf, M. A., Jamal, H. & Whittington, D. (1992) Willingness to pay for water in rural Punjab, Water and Sanitation Report No. 4 (Pakistan and Washington DC: World Bank).

Amigues, J.-P., Bonnieux, F., Le Goffe, P. & Point, P. (1995) *Valorisation des Usages de l'Eau* (Paris: Economica).

De Araujo, J. C. (1996) Cobrança de água bruta no Estado do Ceará. Relatório No. 1, (Fortaleza: Governo do Estado do Ceará/Secretaria dos Recursos Hídricos/Companhia de Gestão dos Recursos Hídricos).

Azevedo, L. G. & Simpson, L. D. (1995) *Management of Water Resources, Economic Notes* (Washington DC: The World Bank).

Barth, F. T. *et al.* (1987) *Modelos para gerenciamento de recursos hídricos* (São Paulo: Coleção ABRH de Recursos Hídricos. Nobel/ABRH).

Baumol, W. & Bradford, D. (1970) Optimal departures from marginal cost pricing, *American Economic Review*, 60.

Biswas, A. K. (1998) Water development and environment, in: A. K. Biswas (Ed.) *Water Resources, Environmental Planning, Management and Development* (Oxford and New York: McGraw Hill).

Braile, P. M. (1979) *Manual de tratamento de águas residuárias industriais* (São Paulo: CETESB).

Cabral, B. (1997) *Direito administrativo, tema água: coletânea de leis relativas a recursos hídricos e meio ambiente* (Brasília: Senado Federal).

Carrera-Fernandez, J. (1992) *Cobrança pelo uso da água de mananciais* (Salvador: HIGESA, (Relatório de consultoria)).

Carrera-Fernandez, J. (1993) *Otimização de usos múltiplos e cobrança pelo uso da água na bacia do Alto Paraguaçu* (Salvador: CRH/CONBEC, Relatório de consultoria).

Carrera-Fernandez, J. (1996) *Projeto de implantação da cobrança pelo uso e poluição da água dos mananciais do Alto Paraguaçu e Itapicuru* (Salvador: SRH/BID, Relatório de consultoria).

Carrera-Fernandez, J. (1997a) Cobrança e preços ótimos pelo uso e poluição das águas de mananciais, *Revista Econômica do Nordeste*, 28(3), pp. 249–277.

Carrera-Fernandez, J. (1997b) *Ampliação do estudo de cobrança pelo uso e poluição da água em corpos d'água do domínio do Estado da Bahia e complementação da regulamentação da lei estadual. Convênio Superintendência de Recursos Hídricos/Banco Mundial (BID)* (Salvador: Relatório de consultoria).

Carrera-Fernandez, J. (1997c) *Usos da água: aspectos econômicos—curso de gestão de recursos hídricos* (Salvador: Centro Interamericano de Recursos da Água (CIRA)/Universidade Católica do Salvador (UCSal)).

Carrera-Fernandez, J. (1997d) *Usos da água: aspectos econômicos: notas de aula para o curso de pós-graduação (lato senso) de gerenciamento e conservação de recursos hídricos* (Salvador: CIRA/UCSal, novembro de).

Carrera-Fernandez, J. (1997e) *Economia dos recursos hídricos* (Salvador: Curso de Mestrado em Economia da UFBA) Texto para discussão.

Carrera-Fernandez, J. (1998) O princípio dos usos múltiplos dos recursos hídricos e o custo social da energia elétrica no Brasil. Anais do III Encontro Regional de Economia da ANPEC, *Revista Econômica do Nordeste*, 28, n. especial, julho, pp. 953–966.

Carrera-Fernandez, J. (1999a) *Estudo de cobrança pelo uso da água na bacia hidrográfica do rio Pirapama. Relatório Final* (Recife: Companhia Pernambucana de Meio Ambiente (CPRH)/Department for International Development (DEFID)/Environment Resources Management (ERM)).

Carrera-Fernandez, J. (1999b) Estudo de cobrança pelo uso da água na bacia hidrográfica do Vaza-Barris. Relatório Final (Brasília, Secretaria de Recursos Hídricos do Ministério do Meio Ambiente (SRH/MMA)).

Carrera-Fernandez, J. (2000a) Research on charging for water use in watersheds: theories, methodologies and an analysis of the studies in Brazil, in: C. Tortajada & A. K. Biswas (Eds) *Water Pricing and Public–Private Partnership in the Americas*, pp. 140–177 (Inter-American Development Bank and National Water Agency of Brazil)

Carrera-Fernandez, J. (2000b) *A valorização da água e a cobrança pelo uso: teoria, metodologias e um estudo de caso para a bacia hidrográfica do rio Pirapama em Pernambuco* (Salvador: CME/UFBA).

Carrera-Fernandez, J. & Gmunder, U. (Orgs.) (1998) *Uso eficiente de recursos naturais e uma política tributária ecológica*, pp. 177–207 (Salvador: Goethe-Institut/ICBA).

Carrera-Fernández, J. & Lima, J. F. (1999) O custo social dos recursos hídricos em bacias hidrográficas internacionais: o caso da bacia do Paraná, *Revista Análise Econômica*, 32, pp. 70–86.

Cedraz, A. (1998) *A lei dos recursos hídricos brasileira. Uso eficiente de recursos naturais e uma política tributária ecológica* (Salvador: Goethe-Institut (ICBA)).

Coase, R. (1960) The problem of social cost, *The Journal of Law and Economics*, 3(1).

COGERH (2000) Internal report, Compauhia de Gestao dos Recursos Hídricos, Fortaleza.

Consórcio Icf Kaiser-Logos (1998) *Cobrança pelo uso da água* (Reprodução Restrita).

Delgado, C. D. & Alberich, M. V. E. (Eds) (1997) *Contribuciones al Manejo de los Recursos Hídricos en America Latina* (Mexico: Universidad Autónoma del Estado de México).

Easter, K. W., Becker, N. & Tsur, Y. (1996) Economic mechanisms for managing water resources: pricing, permits and markets, in: A. K. Biswas (Ed.) *Water Resources: Environmental Planning, Management and Development* (New York: McGraw Hill).

Faria, D. M. C. P. (1995) *Avaliação contingente em projetos de abastecimento de água* (Brasília: Ministério do Planejamento e Orçamento/SPU/IPEA).

Fundação do Desenvolvimento Administrativo (FUNDAP) (1991) Cobrança do uso da água—subsídios para a implantação. Relatório Preliminar, (São Paulo: Convênio DAEE/FUNDAP).

Fundação do Desenvolvimento Administrativo (FUNDAP) (1993) *Cobrança do uso da água. Relatório Final* (São Paulo: Convênio DAEE/FUNDAP).

Fundação Instituto De Pesquisas Econômicas (FIPE) (1993) Elaboração de estudo para implantação da cobrança pelo uso dos recursos hídricos do Estado de São Paulo—proposta técnica, (São Paulo: Convênio FIPE/Departamento de Águas e Energia Elétrica (DAEE)/Consórcio Nacional de Engenheiros Consultores S.A. (CNEC)).

Fundação Instituto De Pesquisas Econômicas (FIPE) (1994a) Elaboração de estudo para implantação da cobrança pelo uso dos recursos hídricos do Estado de São Paulo—definição de hipóteses iniciais e plano geral de trabalho consolidado. Relatório Parcial RP 0 – Rev. 01 (São Paulo: Convênio FIPE/Departamento de Águas e Energia Elétrica (DAEE)/Consórcio Nacional de Engenheiros Consultores S.A. (CNEC)).

Fundação Instituto De Pesquisas Econômicas (FIPE) (1994b) Elaboração de estudo para implantação da cobrança pelo uso dos recursos hídricos do Estado de São Paulo—consolidação dos aspectos relevantes da legislação. Relatório Parcial RP 01 – Rev. 02 (São Paulo: Convênio FIPE/Departamento de Águas e Energia Elétrica (DAEE)/Consórcio Nacional de Engenheiros Consultores S.A. (CNEC)).

Fundação Instituto De Pesquisas Econômicas (FIPE) (1994c) Elaboração de estudo para implantação da cobrança pelo uso dos recursos hídricos do Estado de São Pauloconsolidação dos aspectos relevantes da experiência estrangeira. Relatório Parcial RP 02 – Rev. 01 (São Paulo: Convênio FIPE/Departamento de Águas e Energia Elétrica (DAEE)/Consórcio Nacional de Engenheiros Consultores S.A. (CNEC)).

Fundação Instituto De Pesquisas Econômicas (FIPE) (1995a) Elaboração de estudo para implantação da cobrança pelo uso dos recursos hídricos do Estado de São Paulo—alternativas de modelos gerenciais e jurídicos. Relatório Parcial RP 03 – Rev. 01 (São Paulo: Convênio FIPE/Departamento de Águas e Energia Elétrica (DAEE)/Consórcio Nacional de Engenheiros Consultores S.A. (CNEC)).

Fundação Instituto De Pesquisas Econômicas (FIPE) (1995b) Elaboração de estudo para implantação da cobrança pelo uso dos recursos hídricos do Estado de São Paulo—análise das metodologias de preços médios. Relatório Parcial RP 04 – Rev. 01 (São Paulo: Convênio FIPE/Departamento de Águas e Energia Elétrica (DAEE)/Consórcio Nacional de Engenheiros Consultores S.A. (CNEC)).

Fundação Instituto De Pesquisas Econômicas (FIPE) (1995c) Elaboração de estudo para implantação da cobrança pelo uso dos recursos hídricos do Estado de São Paulo—simulação do potencial de cobrança pelo uso dos recursos hídricos. Relatório Parcial RP 05 – Rev. 01 (São Paulo: Convênio FIPE/Departamento de Águas e Energia Elétrica (DAEE)/Consórcio Nacional de Engenheiros Consultores S.A. (CNEC)).

Fundação Instituto De Pesquisas Econômicas (FIPE) (1995d) Elaboração de estudo para implantação da cobrança pelo uso dos recursos hídricos do Estado de São Paulo—pesquisa de disposição a pagar. Relatório Parcial RP 06 – Rev. 01 (São Paulo: Convênio FIPE/Departamento de Águas e Energia Elétrica (DAEE)/Consórcio Nacional de Engenheiros Consultores S.A. (CNEC)).

Fundação Instituto De Pesquisas Econômicas (FIPE) (1995e) Elaboração de estudo para implantação da cobrança pelo uso dos recursos hídricos do Estado de São Paulo—proposições de alternativas de modelos e instrumentos de implementação. Relatório Parcial RP 07 – Rev. 01 (São Paulo: Convênio FIPE/Departamento de Águas e Energia Elétrica (DAEE)/Consórcio Nacional de Engenheiros Consultores S.A. (CNEC)).

Fundação Instituto De Pesquisas Econômicas (FIPE) (1996) *Elaboração de estudo para implantação da cobrança pelo uso dos recursos hídricos do Estado de São Paulo—consolidação dos resultados do 1°seminário. Relatório Parcial RP 09* (São Paulo: Convênio FIPE/Departamento de Águas e Energia Elétrica (DAEE)/Consórcio Nacional de Engenheiros Consultores S.A. (CNEC)).

Fundação Instituto De Pesquisas Econômicas (FIPE) (1996b) Elaboração de estudo para implantação da cobrança pelo uso dos recursos hídricos do Estado de São Paulo—plano geral de implementação. Relatório Parcial RP 10, (São Paulo: Convênio FIPE/Departamento de Águas e Energia Elétrica (DAEE)/Consórcio Nacional de Engenheiros Consultores S.A. (CNEC)).

Fundação Instituto De Pesquisas Econômicas (FIPE) (1996c) Elaboração de estudo para implantação da cobrança pelo uso dos recursos hídricos do Estado de São Paulo—detalhamento dos programas para implantação da cobrança. Relatório Parcial RP 11 (São Paulo: Convênio FIPE/Departamento de Águas e Energia Elétrica (DAEE)/Consórcio Nacional de Engenheiros Consultores S.A. (CNEC)).

Fundação Instituto De Pesquisas Econômicas (FIPE) (1996d) Elaboração de estudo para implantação da cobrança pelo uso dos recursos hídricos do Estado de São Paulo—consolidação dos aspectos relevantes da experiência estrangeira. Relatório Parcial RP 02 – Rev. 02 (São Paulo: Convênio FIPE/Departamento de Águas e Energia Elétrica (DAEE)/Consórcio Nacional de Engenheiros Consultores S.A. (CNEC)).

Fundação Instituto De Pesquisas Econômicas (FIPE) (1997a) Elaboração de estudo para implantação da cobrança pelo uso dos recursos hídricos do Estado de São Paulo. Relatório Final (São Paulo: Convênio FIPE/Departamento de Águas e Energia Elétrica (DAEE)/Consórcio Nacional de Engenheiros Consultores S.A. (CNEC)).

Fundação Instituto De Pesquisas Econômicas (FIPE) (1997b) Estudo do princípio usuário-pagador nas bacias hidrográficas dos rios Paraíba do Sul e Doce—disposição a pagar na bacia do Paraíba do Sul e Região Metropolitana do Rio de Janeiro. Relatório Final (São Paulo: Convênio FIPE/Departamento Nacional de Águas e Energia Elétrica (DNAEE)).

Fundação Instituto De Pesquisas Econômicas (FIPE) (1997c) Estudo do princípio usuário-pagador nas bacias hidrográficas dos rios Paraíba do Sul e Doce—cálculo da tarifa média e simulações. Relatório Final (São Paulo: Convênio FIPE/Departamento Nacional de Águas e Energia Elétrica (DNAEE)).

Fundação Instituto De Pesquisas Econômicas (FIPE) (1998) Estudo sobre a cobrança pelo uso da água nas bacias hidrográficas dos rios Paraíba do Sul e Doce—Relatório Final (Rio Doce) (São Paulo: Convênio FIPE/Departamento Nacional de Águas e Energia Elétrica (DNAEE)).

Garrido, R. (1993a) *Estudos de cobrança pelo uso da água na bacia de Fêmeas* (Secretaria de Recursos Hídricos Saneamento e Habitação (SRHSH), Superintendência de Recursos Hídricos (SRH) / Governo do Estado da Bahia).

Garrido, R. A. (1993b) Indústria como usuária dos recursos hídricos: notas para discussão na (Cubatão: CIESP).

Garrido, R. (1996) *Contribuição ao Plano Nacional de Recursos Hídricos* (Brasília, DF: Secretaria de Recursos Hídricos do Ministério do Meio Ambiente, dos Recursos Hídricos e da Amazônia Legal).

Garrido, R. (1997) *Aspectos institucionais do planejamento e gestão dos recursos hídricos* (Salvador: Programa CIRA, Universidade Católica do Salvador, v. I e II) (Apostila para o Curso de Gestão e Conservação dos Recursos Hídricos).

Garrido, R. (1998a) *Curso de gestão de recursos hídricos: aspectos institucionais, Volume 1* (Salvador: Universidade Católica do Salvador).

Garrido, R. (1998b) *As bases para a política nacional de recursos hídricos no Brasil. Uso eficiente de recursos naturais e uma política tributária ecológica* (Salvador: Goethe-Institut (ICBA)).

Garrido, R. (2003) *Considerations on Price Setting for Water Use Charge in Brazil* (TWCWM, ANA, IADB). Available at www.iabd.org

Garrido, R. & Carrera-Fernandez, J. (1997) Metodología para la determinación de los precios óptimos y cobro por el uso y contaminación de las cuencas de Paraguaçu e Itapicuru, in: C. D. Delgado & M. V. E. Alberich (Eds) *Contribuciones al manejo de los Recursos Hídricos en America Latina* (Mexico: Universidad Autónoma del Estado de México).

Governo do Estado da Bahia (1997) *Manual de outorga de direito de uso da água* (Salvador: Superintendência de Recursos Hídricos (SRH) / Secretaria de Recursos Hídricos, Saneamento e Habitação (SRHSB)).

Governo do Estado de Pernambuco (1999) *Diagnóstico Ambiental Integrado da Bacia do Pirapama—Projeto Pirapama* (Recife: CPRH).

Greene, W. H. (1997) *Econometric Analysis*, 3rd edn. (New Jersey: Prentice Hall).

Hanemann, W. M. (1984) Welfare evaluation in contingent valuation experiments with discrete responses *American Journal of Agricultural Economics*, 66, pp. 332–342.

Harberger, A. C. (1972) *Project Evaluation: Collected Papers* (Chicago: The University of Chicago Press).

Hearne, R. & Easter, W. K. (1995) Water allocation and water markets: an analysis of gains-from-trade in Chile Technical Paper, No. 15 (Washington DC: World Bank).

Judge, G. G., Griffiths, W. E., Hill, R. C., Lutkepohl, H. & Lee, T. (1985) *The Theory and Practice of Econometrics*, 2nd edn (New York: John Wiley and Sons).

Kelman, J. (1997). Gerenciamento de recursos hídricos: outorga e cobrança. *Anais do XII Simpósio Brasileiro de Recursos Hídricos, Vitória.*

Lanna, E. (1994) Estudos para cobrança pelo uso de água bruta no Estado do Ceará—simulação tarifária para a bacia do rio Curu. Relatório n. 1 (Fortaleza: Governo do Estado do Ceará/Secretaria dos Recursos Hídricos / Projeto de Desenvolvimento Urbano (PROURB) / Companhia de Gestão dos Recursos Hídricos do Ceará (COGERH)).

Lanna, E. (1995a) Estudos para cobrança pelo uso de água bruta no Estado do Ceará—simulação tarifária para a bacia do rio Curu. Relatório n. 2-A (Fortaleza: Governo do Estado do Ceará/Secretaria dos Recursos Hídricos / Projeto de Desenvolvimento Urbano (PROURB) / Companhia de Gestão dos Recursos Hídricos do Ceará (COGERH)).

Lanna, E. (1995b) *Cobrança pelo uso da água: reflexões a respeito de sua aplicação no Brasil* (Porto Alegre: IPH/UFRGS).

Lanna, E. & Pereira, J. S. (1976) *Sacuarema—sistema de apoio a cobrança pelo uso da água e de recursos do meio ambiente* (Porto Alegre: Instituto de Pesquisas Hidráulicas/UFRS).

Lanna, E. & Pereira, J. S. (1996) Simulação da cobrança pelo uso da água na bacia do no dos Sinos, in: *Simposio Italobrasileiro de Engenharia Sanitária e Ambiental*, vol. 1, p. 78 (Brasil: Gramado).

Layard, P. R. G. & Walters, A. A. (1978) *Microeconomic Theory* (New York: McGraw Hill).

Lypsei, R. G. & Lancaster, K. J. (1956–7) The general theory of the second best *Review of Economic Studies*, 24, pp. 11–32.

Loucks, D. P., Stedinger, J. R. & Haith, D. A. (1981) *Water Resources Systems Planning and Analysis* (New Jersey: Prentice Hall).

Lund, H. F. (1980) *Manual para el control de la contaminación industrial* (Madrid: Instituto de Estudios de Administración Local).

Mas-Colell, A. Whinston, M. D. & Green, J. R. (1995) *Microeconomic Theory* (New York: Oxford University Press).

Mitchell, R. C. & Carson, R. (1993) *Using Surveys to Value Public Goods: The Contingent Valuation Method* (Washington DC: Resources for the Future).

da Motta, R. S. (1998a) *Manual para valoração econômica de recursos ambientais* (Brasília: Ministério do Meio Ambiente, dos Recursos Hídricos e da Amazonia Legal).

da Motta, R. S. (1998b) *Utilização de critérios econômicos para a valorização da água no*, Texto para Discussão n. 556 (Brasil: IPEA).

Nowak, F. (1995) *Le Prix de l'Eau* (Paris: Economica).

Pompeu, C. T. (1996) O projeto de lei brasileiro para o setor de recursos hídricos, *Anais do XI Seminário—Curso do Programa CIRA* (Salvador: Universidade Católica do Salvador).

République Française (1995) *Pollution des Eaux: Redevances. Direction des Journaux Officiels* (Paris).

Rhodes, Jr., G. & Sampath, R. K. (1988) Efficiency, equity and cost recovery implications of water pricing and allocation schemes in developing countries, *Canadian Journal of Agricultural Economics*, 36, pp. 103–117.

Rosegrant, M.W. & Binswranger, H.P. (1993) Markets in tradable water rights: potential for efficiency gains in developing country irrigation. Draft Paper (Washington DC).

Secretaria de Planejamento, Orçamento e Coordenação da Presidência da República (SEPLAN-PR) (1994) *Projeto Aridas—uma estratégia de desenvolvimento sustentável para o Nordeste* (Brasília: SEPLAN-PR).

de Souza, M. P. (1995) A cobrança e a água como bem comum, *Revista Brasileira de Engenharia—Caderno de Recursos Hídricos*, 13(1).

Sunman, H. (1992) The application of charging schemes for the management of water pollution: experience and prospects. Draft Paper.

Simpson, L.D. (1996) Cost recovery for water resource projects. Draft Paper (Washington DC).

Varian, H. R. (1978) *Microeconomic Analysis* (New York: Norton Company Inc).

Warner, R. (1995) *Water Pricing and the Marginal Cost of Water* (Melbourne: Water Services Association of Australia).

Water Charges: Paying for the Commons in Brazil

BENEDITO P. F. BRAGA, CLARICE STRAUSS & FATIMA PAIVA
Agencia Nacional de Aguas, Brasilia DF, Brazil

Preventing the Tragedy

While writing *The Tragedy of the Commons* during the late 1960s, Garrett Hardin (1968) could hardly have realized the tremendous impact that his article, based on a pamphlet written in 1833 by William Foster, would have in the world's embryonic environmental debate. In fact, it has become a classical reference whenever the issue of the common goods is raised. It seemed particularly useful to bring Hardin's metaphor to the discussion concerning water charges in Brazil, since according to Brazilian legislation, water bodies are a public good. Water is, in fact, a 'common', accessible to any person in the country. Water bodies cannot become private property; they are under either the federal government or the state's jurisdiction, depending on whether or not they cross or serve as state boundaries.

Garret Hardin's *Tragedy of the Commons* is an illustration of the incorrect use of public resources by private interests when individuals, with free and equal rights to access the commons, acting separately in an attempt to maximize their own utilities, may collectively over-exploit and deplete natural scarce resources. Hardin asserts that "freedom in a commons brings ruin to all"; in fact, it supports by omission, the over-utilization of natural scarce resources, and ultimately the complete depletion of the resource. It must be emphasized that the freedom of the commons becomes a tragedy only by the time the carrying capacity of the system reaches its limit, so it is system sensitive. At this point, freedom must be restricted, somehow, by somebody.

The application of Hardin's metaphor to water resources today requires no further explanation. Water, is a scarce resource essential to all forms of life in the planet.

For this reason, it is defined in the Brazilian Constitution as a public good and in the Water Law 9433/97 as an economic good. Some kind of exclusion mechanism must be enforced in order to prevent, or repair, different degrees of depletion of the water bodies. The enormous conflicts over water resources must also be arbitrated, whether they relate to a conflicting category of uses or over a fair distribution of the resource among different users.

There is no technical solution that would ensure the rational, equitable and sustainable use of the water resources. Technical and economic instruments are necessary. The complexity of the task determines that it is necessary for the solution to be a political one. Engagement and co-operation are required from all stakeholders. Changes in values are required at different levels, stirring all stakeholders: public sector, private sector and civil society alike. Drastic institutional adaptations are called for in order to respond to the new values and to address the issue at different levels with appropriate strategies.

Accordingly, the Brazilian 1997 Water Law, which establishes the National Water Policy, poses a new paradigm on water management. It changes the historical Brazilian paradigm on public administration, by promoting active partnerships and co-responsibility over the water management by all relevant stakeholders. Decisions are decentralized, river basin committees are established by representatives of the public sector, private sector and civil society alike. These committees are empowered with real responsibilities over the basin's water management, including decisions with regard to water allocation among different users and water charges.

Water Charges

The Brazilian Constitution establishes that water is a public good, under the jurisdiction of either the state or the Union. A given water body is under the jurisdiction of the Union whenever it flows or serves as a boundary to more than one state. However, the hydroelectric potential and the water in reservoirs built by the Federal Government are also under the jurisdiction of the Union, no matter whether the river belongs to the Union or to the state.

Distinctions: Charges vs. Tariffs

In order to clarify concepts, a distinction between charges and tariffs needs to be made at this stage. Tariffs relate to the provision of services such as potable water and sewerage; they intend to cover the operational and maintenance costs of such services and to provide an economic return to the invested capital, in the case of privately owned enterprises. On the other hand, charges do not relate to services or investments. Charges relate to the use or to the rights to use the natural resource itself: bulk water. Charges could be seen as an economic return to a natural capital and also as an incentive to rational use. Charges are an economic compensation for using a share of the water 'commons'. This paper will deal with water charges. The debate over water tariffs is beyond the scope of this study.

Legal Framework

Historical background. The legal basis for water charges in Brazil is rather old. It was laid down by the 1916 Civil Code, which established that the use of public goods (commons) could be either free or subject to charges, in accordance with the applicable Union, states and municipal laws. Accordingly, the 1934 Water Code, the 24.642/34 Act, established that the use of the water commons could be either charged or free, according to the laws and the regulations of the federative unit endowed with the water jurisdiction.

Later, the 1981 National Environment Policy, through the 6938/81 Act, determined that the polluters had an obligation to protect the environment and/or to compensate for the damage inflicted by them on the environment. It also established that the user would pay for the exploitation of environmental resources with economic purposes. Finally, the 9433/97 Water Law established water charges (abstraction and effluent discharges) as one of the instruments for water resources management. The 9984/2000 Law attributed to ANA the responsibility to enforce charges over water resources use under the Union's jurisdiction. The 'shared responsibilities' established by the 9433/97 and the 9984/2000 Acts themselves generated a revolution on the public management paradigm. Things had to be done differently. Nobody really knew how. The 'how' had to be designed by all stakeholders, requiring the engagement of civic society, private sector and different spheres of government altogether.

The national water resources policy (NWRP). It is the Union's task to implement the National Water Resources Management System and to dictate the national criteria for bulk water charges to be followed by the states. The 9433/97 Act established the current Brazilian National Water Resources Policy principles. The principles of the law as well as the instruments established by it are well tailored to address the potential and current water conflicts, resulting from the growing water demand from urban, industrial and agricultural growth. Water charges as established by the NWRP are a water management instrument. The National Water Agency, ANA, was created through the 9984/2000 Act, with the mandate to implement the NWRP and to serve as a regulatory agency for different water uses in the country.

The NWRP aims to ensure present and future generations have a water supply that meets both the quality and quantity required for the different uses. It aims at promoting a rational and integrated use of the water resources, including navigation, in order to ensure sustainable development. Finally, it aims at prevention and defence from water-related critical events, both natural and the ones originating from inadequate human interventions.

The NWRP principles are as follows:

- Water is a common public good.
- Water is a natural resource with economic value.
- In situations of critical scarcity, priority of use will be given to human water supply and animal consumption.
- Water management must always make provision for the multiple uses of water.
- The river basin is the basic territorial unit for the planning and implementation of the NWRP.
- Water management must be decentralized. Decision making is required to be a participative process that includes government representatives as well as private sector and community representatives.

NWRP promotes a systematic water management system that integrates quality and quantity as inseparable aspects of the water resources. It takes into account the country's regional diversity in relation to its specific physical, demographic, biotic, economic, social and cultural qualities. It establishes a decentralized integrated management system at the river basin level, based on partnerships between legitimate representatives of all stakeholders: government, private sector and civil society (ANA, 2002).

Main challenges imposed by the NWRP. These principles pose new challenges in several spheres of action, including:

- On the administrative sphere, it requires that the river basin be the management unit. As a consequence, new institutional arrangements are required, since it is not part of the existing administrative division of the country. Consequences range from databases to the establishment of new managerial authorities.
- On the social side, it poses a whole new paradigm of shared management, where legitimate civil society and private sector representatives are empowered as river basin authorities along with the public sector, forming the river basin committee.
- On the economic side, it establishes that water has an economic value, and therefore charges are made to pay for its use.
- On the technical side, a major challenge arises as information bases and monitoring systems need to be created, along with dynamic rosters of users, in order to enforce the new management responsibilities.

Institutional Framework

The new water management paradigm and, specifically, the process of quantifying water charges, and enforcing them, entails an institutional framework which includes river basin committees, river Basin Agencies, the National Council for Water Resources (CNRH), the State Water Resources Councils (CERH), the National Water Agency (ANA) and State Water Authorities. The main responsibilities of these institutions regarding water charges are as follows:

(i) The river basin committee suggests the values to be charged for the different uses. The committee also approves and supervises the execution of the river basin water resources plans, which lay down the priorities and criteria for the management of the basin's water resources. The river basin committee, a decision-making forum, is composed of democratically elected representatives among stakeholders in the basin, from the public sector (including municipal and state level water authorities), private sector and civil society. Therefore, it is political in nature. The river basin agencies provide technical support to the river basin committee. Under this mandate, the basin agency provides studies on unit rates for water charges for different water uses in the basin, and elaborates the river basin water resources plan.

(ii) The National Council for Water Resources (CNRH) establishes additional guidelines for the enforcement of the legal instruments, as well as general criteria for water charges. In addition, the CNRH approves the proposal and in co-ordination with the basin committee defines the investment priorities for the resources levied at basin level.

(iii) The National Water Agency (ANA) provides technical support to river basin committees and to the National Water Resources Council (CNRH) in the definition of the unit rates for water charges in different river basins in the country. ANA is also responsible for the collection of charges in federal rivers (rivers shared by two or more states).

In short, the river basin agencies propose the amounts to the committees. The committees, in turn, establish the water charges procedures and suggest, to the national authority, the CNRH, the values to be charged. Furthermore, the 9984/2000 Act established that ANA, in co-ordination with the committees and the water agencies, should carry out the research required to define the optimum charges. The proposal is then submitted to the CNRH, which has the final word on the values to be charged. The CNRH also defines, in co-ordination with the committees, the investment priorities for the water charges revenues.

'Shared Management': A New Paradigm

This new institutional framework poses several new challenges to the water management issue. The main challenge is the implicit new paradigm on governance, where civil society and private sector representatives are empowered at the river basin level, as members of the river basin committee, with responsibilities that used to be exclusive to the public sector.

These challenges could be divided, according to their nature, as:

(1) Political challenges related to modifications in the decision-making power structures. Under the new management paradigm, governance is ensured through a fine tuned 'social-engineering' which requires effective partnerships between the public sector, civil society and the private sector. This requirement was laid down by the 9433/97 and 9984/2000 Acts, where the principles and mechanisms for integrated, decentralized and participatory management were established.

(2) Legal challenges dealing with the necessary legal modifications to the new management paradigm, including coping with legal omissions, attuning the water related legislation from the different states included in the basin within them with the national legislation, as well as harmonizing the multiple sectors specific legislation.

(3) Technical and economical challenges. On the technical side, the new management structure requires information and follow-up systems, which themselves are a great challenge. The technical viewpoint is addressed from the establishment of a database for users to the improvement of the water related environmental information and monitoring system that would support the enforcement of the law.

Defining the Bulk Water Charges

According to the 9433/97 Act, the bulk water charges have three objectives: (1) to acknowledge water as an economic good; (2) to promote its rational use; and (3) to raise funds for the required actions at the river basins. A fraction of the water charges (7.5%) may be used to cover the operational costs of the National Water Resources Management System's agencies. The following parameters should be considered, among others, as inputs for the definition of the values to be charged:

- For water abstraction: the volume diverted and its flow variation.
- For water effluents: the volume discharged, the flow variation and the biogeochemical characteristics of the effluent.

The 9433/97 Act also establishes that the uses subject to water permits will also be subject to charges. In that sense, water uses that do not include extraction, consumption or discharges of effluents in the water bodies would be subject to specific procedures in order to define the charges to be paid by them. This is the case of small hydropower plants with installed capacity lower than 30 MW.

Hydroelectric power plants with a generating capacity greater than 30 MW must pay financial compensation for the exploitation of water resources. The total value due by them is the equivalent to 6.75% of the value of the energy produced by the plant. A small percentage of these charges, 0.75%, is allocated to the Ministry of Environment, as payment for water use, as stated in the 9433/97 Act. Currently, small plants with power capacity lower than 30 MW do not pay financial compensation. However, the same legal Act makes it possible for other formulations to be established to levy these small hydropower plants.

Ceara State Experience

Historical Background

The Ceara State Water Resources Policy and its Management System were established in 1992, through the state 11.996/92 Act. It established water charges in the terms agreed by the State Council on Water Resources (CERH). From 1992, Ceara state moved steadily towards the implementation of water permits, water charges, water information systems and water resources basin plans, promoting the necessary institutional adaptations in order to enforce the state's water policy.

By October 1996, CERH established that the Water Resources State Agency (COGERH) would be responsible for supplying bulk water from the reservoirs and other water bodies. The same legal Act authorized COGERH to charge for the use of water bodies under its control. This was the first experience of bulk water charges in Brazil (COGERH, 2002).

A value of R$0.60/m^3 was set for the industry (US1.00 = R3.00). This value corresponded to 50% of the tariffs established by the State Water Supply and Sanitation Company (CAGECE) for treated water for industrial users abstracting more than 70 m^3/month. A value of R$0.01/m^3, which is equivalent to 1/60 of the tariff established by the CAGECE for industrial use, was established for the urban water supply concessions. Afterwards, aiming at establishing a global policy of water charges in the state, CERH, through the 003/97 Deliberation, approved the following general criteria:

(1) *Industries*. Value = 50% of the CAGEGE tariff for industrial use with consumption higher than 70 m^3/month.
(2) *Urban water supply*. Value = 1/60 of the CAGECE tariff for industrial use with consumption higher than 70 m^3/month.
(3) *Users of pressurized systems or of open channel systems*. Value to be fixed for each system by the State Water Authority (except users related in (1).
(4) *Irrigation, fishing and aquaculture*. Value higher than or equal to 1/600 of the value charged for industrial users, to be fixed by the basin committee, when applicable.

The state's bulk water management steadily developed, disconnected from the Sewer and Water Supply Policy. Furthermore, a growing demand for new industrial projects was observed, which meant that there was a need for new investments in water supply infrastructure. Taking these two factors into account, the values charged for urban water supply and industrial use were raised to R\$0.012/m^3 and R\$0.67/m^3 respectively.

In 2000, the value for urban consumption in the Integrated System of the Fortaleza Metropolitan Area was fixed at R\$0.028/ m^3. The purpose of this increase was to include the pump stations energy costs in the charges. Currently, the State Water Authority is carrying out additional studies for defining and implementing a more integrated water charges policy to attain full cost recovery.

A Quasi-water Market Experience in the Jaguaribe River

In 2001, Brazil's first experience in a sort of temporary water market in the Jaguaribe river basin was established in Ceara state (Figure 1). Although technically it cannot be called as such because water is a public good, ANA and the state authorities implemented a procedure through which farmers irrigating crops of low aggregated value (LAV) gave up their rights to farmers irrigating crops of high aggregated value (HAV). The process involved a compensation mechanism to the LAV crop farmers and payment by the HAV crop farmers.

In order to address this problem, and after intense negotiations, the farmers agreed to pay R\$0.01 for every cubic metre of water they used. The funds raised through these water charges were used to compensate the farmers that were left without enough water to irrigate their crops, provided that they were willing to change, in the near future, water intensive crops, such as rice, for less water intensive ones with higher value added. The farmers that joined the programme could also access credit to buy new agricultural equipment as well as participate in training programmes.

This was an innovative and pioneering water charges experience. Stakeholders realized the need to use ad hoc excluding mechanisms in order to prevent a 'complete ruin to all', associated with a severe scarcity of water. The charges were voluntarily accepted as the best strategy to deal with the severe scarcity. Compensation mechanisms were agreed upon. Stakeholders stopped acting as individual 'profit-maximizers' to be part of a community, part of a natural system where the access to the 'commons' needed to be regulated for the common good.

Although this experience may be seen as being very close to a real water market experience, it cannot be stated to be one as such because water is a public good in accordance with the Brazilian Constitution. As a consequence, it cannot be privately owned nor sold. Therefore, a stakeholder who obtains a given water permit may use it, or not, but he or she is not entitled to sell their rights to a third party. The quasi-market experience is then even more valuable, since it relies on true engagement and collaboration from all stakeholders in transforming a crisis in which all would be ruined into a 'win–win' situation.

Farmers who voluntarily refrained from using the water quota they were entitled to, and informed the water authorities, could have their water reallocated to other farmers. The reallocation procedure was carried out giving consideration to farmers whose crops had greater value added. Farmers giving up their 'water rights' were compensated for their loss. They were also entitled to obtain training courses aimed at identifying less water intensive and more profitable crops. In addition, they could access credits that would support investments needed to produce the required production shift. During the process, 59 million/ m³ of water (about 5.70 m³/s that had been traditionally used by rice crops) was transported to more productive uses. This was made possible with the release of 60% or less of the water volume stored in the reservoirs and the release of roughly 15% of water demanded by the rice crops. In addition, the project ensured: (1) 100% of the urban supply demand; (2) 100% of the prawn cultivation demand; (3) 120% of the water demanded by the fruit crops. The financial balance included around US$2 million for costs, and US$15 million of economic benefits resulted from sales of the produced goods; this corresponded

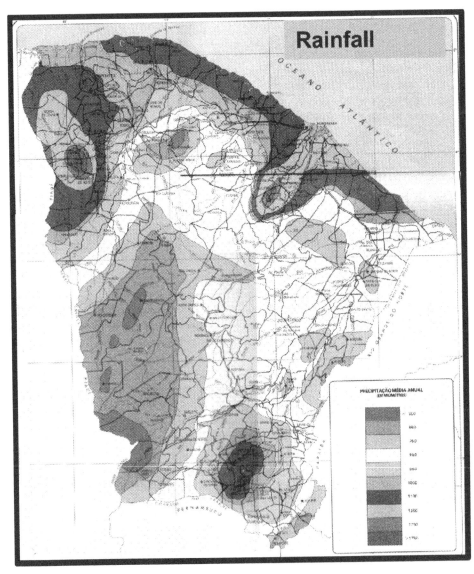

Figure 1. The Jaguaribe River Basin quasi-water market experience.

to a revenue increase of about 468% when compared with the outcomes of previous years. As all this occurred during a water scarce year, so the results are even more encouraging.

The development objective behind water charges was in fact to prepare this semi-arid region so it could focus on higher value added agricultural crops that are less water intensive. This experience is particularly relevant since many developed countries had their farmers paying for the water use only very recently.

Water Charges in Paraiba do Sul River Basin

Implementing Water Charges at a National River Basin

The Paraiba do Sul river basin is located in the south-east region, Brazil's most dynamic economic region (Figure 2). With approximately 8500 industries located in the basin and over 71 000 ha of irrigated agriculture, the Paraiba do Sul river basin is estimated to be responsible for 13% of the country's GDP. Its catchment area is approximately

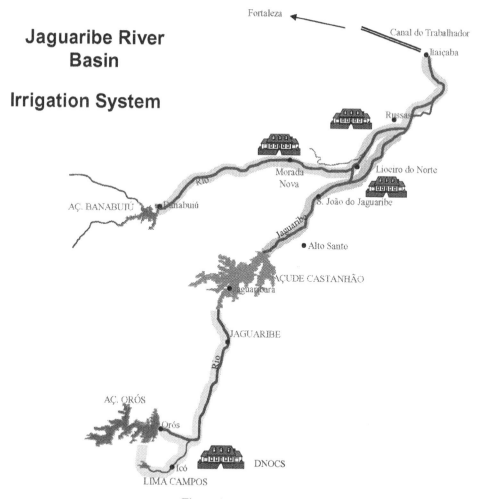

Figure 1. (*continued*).

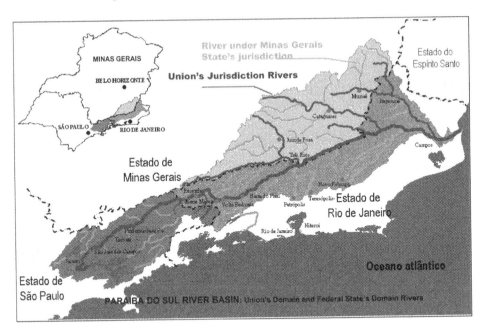

Figure 2. The Paraíba do Sul River Basin and the sharing states of Sao Paulo, Rio de Janeiro and Minas Gerais.

$55\,000\,\text{km}^2$, divided into three states: 25% of its area is under the jurisdiction of the state of São Paulo ($13\,900\,\text{km}^2$), 37.3% under the state of Minas Gerais ($20\,700\,\text{km}^2$), and 37.7% is under Rio de Janeiro's state ($20\,900\,\text{km}^2$). The basin's catchment thus comprises water bodies under the jurisdiction of the Union as well as that of the three states.

The basin's population is approximately 5.4 million inhabitants, 87% of which live in urban areas. The basin supplies water for approximately 14 million inhabitants, more than double the basin's population. Thanks to an inter-basin water transfer, it provides a proportion of the water supply for the 8.7 million inhabitants of the Rio de Janeiro Metropolitan Area. In fact, $160\,\text{m}^3/\text{s}$ is diverted from the Paraiba do Sul to the Guandu river for this purpose. Apart from supplying water for important urban centres, the basin's hydroelectric potential is responsible for over 1.7% of the country's potential.

The high rate of urbanization in the basin and the intensity of its economic activities, particularly industrial activities, have led to an elevated degree of environmental depletion, with an emphasis on water pollution problems, the main causes being untreated urban sewage and industrial effluents as well as inappropriate solid waste disposal. Soil erosion is also a relevant problem in the basin.

The most significant water uses in the Paraiba do Sul are as follows:

- *Urban water supply.* Estimated population: 14 300 000 inhabitants—water supply for domestic use of $65\,\text{m}^3/\text{s}$.
- *Hydroelectric generation.* Installed power: 1500 MW with planned expansion to 2300 MW.
- *Industry.* About 2500 users in São Paulo, 4000 in Rio de Janeiro, and 2000 in Minas Gerais (this includes users connected to the water supply and sanitation public network).

- *Domestic effluents.* The basin receives approximately 249 tons BOD (Biochemical Oxygen Demand) per day from domestic effluents.
- *Rural properties.* Approximately 6000 properties with 71 000 ha of irrigated land.
- *Recreation and tourism.* Canoeing, rafting and fishing take place, mostly in the hilly regions endowed with waterfalls.
- *Aquatic fauna.* 169 species of fresh water fish (115), salt water (38) and exotic species (16).

Implementation of Water Charges

The Paraíba do Sul River Basin Integration Committee (CEIVAP) was established in 1997. Among its responsibilities, the committee analyses and approves the Basin's Investment Plan and proposes the water charges to be implemented in the basin to CNRH. The methodology for water charges was agreed upon after intense debates and negotiations with the representatives of all stakeholders. It was based on scenario studies aimed at matching the basin's revenue expectations intending to raise the necessary funds for implementing the basin's investment programmes. The methodology, the associated unit values and the expected revenues were approved by CEIVAP on 6 December 2001. It was then submitted to, and approved by, the National Water Council (CNRH) on 3 March 2002. By December 2002, CEIVAP's Deliberation 15 and CNRH 27 resolution approved the water charges for both the agricultural sector and for the small hydropower sector.

By March 2003, the National Water Agency (ANA) in co-ordination with CEIVAP and the State Water Authorities, completed the first phase of the self-enrolment process for the basin's water users. Currently, the data are being analysed and water user permits are being granted. The first water charges invoices were distributed to the users and the first payments were made by March 2003. The Basin Agency was approved by CEIVAP's Deliberation 05/2001 and is currently fully operational.

Methodology

The 08/2001 CEIVAP Deliberation approved a methodological proposal for the sewerage and water supply and the industrial sectors. The same methodology was adapted and further approved for the agricultural and animal husbandry sector through the 15/2002 CEIVAP Deliberation. This Deliberation also established a formula for small hydropower water charges, based on the amount of energy being generated. Both Deliberations were further ratified by the National Council for Water Resources (CNRH).

The methodology calculates the monthly values for consumptive and effluent dilution uses, in the Paraíba do Sul Basin taking the following three factors into account:

- volume of diverted water;
- volume of effectively consumed water; and
- volume of water required to dilute the effluents.

The charges were defined on a monthly basis as follows (ANA/Fundação COPPETEC, 2002):

$$C = \underbrace{Q_{cap}.K_0.PPU}_{diversion} + \underbrace{Q_{cap}.K_1.PPU}_{consumption} + \underbrace{Q_{cap}.(1 - K_1)x(1 - K_2.K_3).PPU}_{effluent\ dilution\ (BOD)} \qquad (1)$$

where:

Q_{cap} = volume of abstracted water (m^3/month);

K_0 = multiplying factor to be applied to the abstracted water unit price; (the initial for K_0 = 0.50);

K_1 = consumption coefficient for the activity (the percentage of the abstracted water which is not returned to the river);

K_2 = fraction of the treated effluent in relation to the total volume of effluents produced;

K_3 = efficiency level of BOD reduction (Biochemical Oxygen Demand) in the Effluent Treatment Station;

PPU = Unitary Public Price for extracted volume of water for either consumption or effluent dilution (at R$0.02/m^3).

Charges for the Agricultural Sector

The need for specific criteria for the agricultural sector was acknowledged. For this purpose, simulations were carried out, aimed at evaluating both the potential revenues and the economic impact of water charges over these sectors. Two representative crops, rice and sugar cane, were selected in order to assess the impacts on this sector.

Table 1 presents the coefficients and criteria adopted for the calculation of the crops' water consumption. The same formula in (1) was further approved by CEIVAP for the agricultural sector with the required adaptations to the PPU. During the first stage, the inadequacy of the data and the complexity of the process, particularly the diffuse nature of the agricultural effluents, or the non-point BOD load, resulted in overlooking the agricultural sector's pollution charges.

The application of this methodology resulted in charges of R$331.78 /year/ha for rice crops and of R$260.70/year/ha for sugar cane. The impact of these charges would have been a 17% increase over rice production costs and of 12% over the sugar cane, as shown in Table 2.

Impacts of such magnitude would not be acceptable, since rice crops would become unprofitable. Simulations indicated that the impact of water charges over the State Water Utilities in the basin reached 2.9% over the minimum tariffs only. International experience shows that the agricultural sector does not usually pay for bulk water. However, whenever they pay, they pay less than the other sectors, and they normally participate in the process only some years after the other users do.

Adjustments were required to enforce water charges in the agricultural sector without producing unbearable financial impacts. A reduction in the PPU was suggested. It was assumed that the impact in the sector should be at most 1% of the production costs. The outcome of this criteria was a reduction of 95% of the PPU value, resulting in

Table 1. Water consumption for rice and sugar cane crops

Rice	Sugar cane
Cycle: 120 days	Cycle: 270 days
Consumption : 2.0 l/s/ha = 20.736 m^3/year/ha	Consumption: 0.71 l/s/ha = 16 500 m^3/year/ha
Efficiency: 40%	Efficiency: 39%
BOD release: none	BOD release: none

Source: ANA/ Fundaçãp COPPETEC (2002).

Table 2. Impact of water charges on rice and sugar cane production costs: Simulation 1

Crop	CEIVAP formula					Charges	
	K_0	K_1	K_2	K_3	PPU	(R$/year/ha)	Impact
Rice	0.4	0.4	1	1	0.02	331.78	17.28%
Sugar cane	0.4	0.39	1	1	0.02	260.70	12.59%

Source: ANA/ Fundação COPPETEC (2002).

the following values for the water charges: rice = R$16.59/year/ha and sugar cane = R$13.03 /year/ha. The impacts of these charges in the production costs would be of 0.86% on the rice crops and 0.60% on the sugar cane, as shown in Table 3.

The final PPU values for the agricultural sector were defined as R$0.0004/m^3, with $K_0 = 0.40$ and $K_1 = 0$. Theses values were established by the CEIVAP Deliberation 15/2002 after discussions were carried out on the light of the simulations of economic impacts, and agreements were reached between CEIVAP, ANA, the states representatives and agricultural sector representatives in the basin.

Small Hydropower Sector

For the small hydropower (SHP) the approved water charges were established as follows:

$$C = GH \times TAR \times P$$

where:

C = total value due by each SHP (R$/month)
GH = Generated energy (MWh per month)
TAR = the value of the Energy Tariff established by a Resolution of the National Electric Energy Agency (ANEEL)
P = percentage defined by the CEIVAP currently equal to 0.75%

Self-enrolment

The self-enrolment process co-ordinated by ANA presented the following indicators by March 2003. Table 4 shows the number of enrolled users, which may be either individual users or enterprises, such as the Water Supply State Company of São Paulo (SABESP) responsible for the water supply of several municipalities. The 69 enrolled sanitation users are in fact serving 137 municipalities; each of these users is therefore responsible for several water abstraction and discharge points. Regarding the animal demand, it is observed that of 1939 users only two would pay for bulk water. The others, with

Table 3. Impact of water charges on rice and sugar cane production costs: Simulation 2

Crop	CEIVAP formula					Charges	
	K_0	K_1	K_2	K_3	PPU	(R$/year/ha)	Impact
Rice	0.4	0.40	1	1	0.001	16.59	0.86%
Sugar cane	0.4	0.39	1	1	0.001	13.03	0.60%

Source: ANA/ Fundação COPPETEC (2002).

Table 4. Water users, Paraíba do Sul River Basin

Users	States			Total	
	SP	MG	RJ	Quant (Enrolled users)	%
Industry	113	129	146	388	12.29
Water Supply and Sanitation Service	26	22	21	69	2.19
Irrigation	406	97	37	540	17.11
Animal demand	1616	261	62	1939	61.44
Small hydropower plants	1	5	2	8	0.25
Other users	121	56	35	212	6.72
Total				3156	100

discharges smaller than 1.0 l/s, are considered as being among the insignificant uses and therefore not subject to charges as per the 9433/97 Act determinations.

Conclusions

The implementation of the National Water Resources Management System in Brazil is well underway. A complex system has been established in which the federal, state and municipal governments participate in the decision-making process together with water users and the organized civil society. The river basin has been adopted as the territorial unit for planning and management.

The basic instruments for regulating water use in Brazil are water permits, water charges and control mechanisms. The emphasis given by the National Water Agency in order to foster optimal water uses and effluent discharge treatment has been the use of economic management instruments. Traditional command and control mechanisms have proved to be inefficient. Although control mechanisms are used to a certain extent, they are well balanced with negotiation mechanisms, engaging all relevant stakeholders, thus effectively regulating and minimizing water conflicts.

The results from the intense work in the Paraíba river basin are extremely positive. A bulk water charging formula was implemented in which the user-pay and polluter-pay principle was put into practice. Through a differential valuation scheme it was possible to include all water users in the charging process. The agricultural sector that, throughout the world, is traditionally reluctant to participate in charging schemes, has agreed to pay in the Paraíba do Sul river basin. The key for success seems to be the participation of all stakeholders in the decision process in a bottom-up approach.

References

ANA/Fundação COPPETEC (2002) *Relatório Final Cobrança pelo uso da água bruta na bacia do rio Paraíba do Sul: da a aprovação à preparação para sua aplicação.* Final report (Rio de Janeiro: Fundação COPPETEC).

ANA (2002) *The Evolution of the Organization and Implementation of Water Basin Management in Brazil* (Brasilia: Agência Nacional de Águas).

Companhia de Gestão dos Recursos Hídricos (COGERH) (2002) *A Gestão das Àguas no Ceará* (Fortaleza, Brazil: COGERH).

Hardin, G. (1968) The Tragedy of the Commons, *Science*, 162, pp. 1243–1248.

Water Reform across the State/Society Divide: The Case of Ceará, Brazil

MARIA CARMEN LEMOS* & JOÃO LÚCIO FARIAS DE OLIVEIRA**
*University of Michigan, Ann Arbor, MI, USA
**Departamento Nacional de Obras e Combate às Secas (DNOCS), Fortaleza, CE, Brazil

Introduction

The ubiquitous combination of population growth and limited availability of resources has challenged policy makers and scientists to look for solutions that simultaneously address needs for enhanced service delivery and for environmental sustainability. Within the water sector, public private partnerships (PPPs) have been hailed both as a potential solution to help close the gap between service demand and delivery and as an innovative policy tool to promote positive values, including decentralization, accountability, transparency and public participation. While traditionally PPPs have been defined as contractual agreements between a public agency (federal, state or local) and a for-profit corporation,[1] other forms of partnerships have emerged between governments and various non-profit private partners, such as non-governmental organizations (NGOs), private voluntary organizations (PVOs) and communities. The expansion of PPPs has been especially evident in the case of the delivery of water-related services to the poor where incentives for for-profit private business to participate might be low or absent (Lemos *et al.*, 2002). To reflect such partnerships, new PPP conceptualizations encompass not only the established government-private business partners but also a wider range of actors and approaches combining public and private resources. For example, the UNDP Program for PPPs for the Urban Environment defines these partnerships "as a spectrum of possible relationships between public and private actors for the co-operative provision of infrastructure services. The 'right' relationship is the one that best meets the needs of the partners in the local context. One size does not fit all" (BDP, 2002).

Despite being broad in scope and characteristics, many PPP models share an emphasis on the need for community participation in the formation and implementation of partnerships across the public-private divide. However, while calls for public participation proliferate, there is growing evidence that achievement of meaningful participation is complex and not easily accomplished (Marañon-Pimentel & Wester, 2000; Nickson & Vargas, 2002). One important factor affecting the success of public participation in PPPs is the widespread lack of trust in governments pervading many policy systems, especially those of less developed countries. This lack of trust is informed by decades of broken promises and the endurance of authoritarian practices such as clientelism, corruption, co-optation, and paternalism in public policy making. Whereas until recently, the idea of actors bridging the public-private divide was met with mostly scepticism, increasingly, a cautiously optimistic view of the ability of state and societal actors to form partnerships is gaining terrain, especially in young democracies such as Brazil. Although accounts of policy failure and mistrust still outnumber cases of positive interaction, examples of mutually beneficial partnerships, especially at the local level, accumulate. (For some examples of positive interactions in environment-related policy making, see Watson (1995); Lemos (1998); Lemos & Looye (2003).)

This study aims at contributing to this burgeoning field by examining the implementation of a new model of water management in Ceará, Northeast Brazil.[2] In this drought-ridden region, water related policy making has historically occupied the highest priority in the governmental and societal policy agendas. The study specifically analyses the creation of Users' Commissions by the Ceará Water Resources Management Company (Companhia de Gestão de Recursos Hídricos—COGERH) in two state valleys: the Lower Jaguaribe/Banabuiú and the Curú.[3] Such Commissions are formed by representatives of water users and society as well as state, federal and municipal governments.[4] While traditionally, water management in Ceará has been characterized by centralization, clientelism, and the dominance of technocratic decision making, in a remarkable departure from 'business as usual' the new model implemented by COGERH included the organization of Users' Commissions to debate and decide on the use and allocation of bulk water among different users. The implementation of this model in Ceará is part of a broader initiative to reform the management of water resources in Brazil.[5] The water reform includes the creation of new institutions of management at the national and state levels that emphasize decentralization, public participation, integration and the creation of a water use permit and charge system. Ceará was one of the first states in the nation to reform its system (even before the enactment of the national law) and after almost a decade of active stakeholder participation at the river basin level, its model is heralded as a case of successful implementation of water reform (ANA, 2002).[6] It is also showcased as an example to be pursued by other states.

How did one of the most authoritarian and clientelist policy-making systems in Brazil get transformed into an example of decentralization and successful public participation? What factors account for the positive implementation of water management reform in Ceará? It is argued that in this highly interactive model of water management, COGERH and the Users' Commission have forged partnerships in Ceará in which they share not only goals, but also the responsibility for water use, allocation and conservation. Such partnerships go beyond complementarity to create synergy, which implies that "civic engagement strengthens state institutions and effective state institutions create an environment in which civic engagement is more likely to thrive" (Putnam, 1993, cited by Evans, 1996).

Indeed, the interaction between COGERH and Users' Commissions is likely not only to strengthen both organizations vis-à-vis other groups inside and outside the state but also to forge a model of water management that actively promotes democratic values such as decentralization, accountability, transparency, and public participation. Even if recent empirical evidence suggest that decision making within the Users' Commissions is mostly dominated by large water users, it has also shown that the Commission has created the opportunity for previously disenfranchised groups (e.g. small farms) to participate in water management (Ballestero, 2004). The interaction between COGERH and Users' Commissions may also substantially contribute to a positive policy outcome since public participation can improve water management in several ways. First, it decentralizes policy making by moving the main arena of decision making from the state to the river basin level. Second, it can improve water allocation and use by providing stakeholders with vital information that allows Users' Commissions to make decisions based not only on their immediate needs but also by taking into consideration the sustainability of the whole river basin system. Third, it can increase compliance with water allotments by making users accountable to the decisions carried out by their legitimate representative bodies. This increase in compliance can happen both as a result of users taking responsibility for the decisions they voluntarily agree to as well as a result of the commitment of users to hold each other accountable. Fourth, it promotes widespread awareness of water resources scarcity and the actions needed to guarantee long term system sustainability. Finally, after close to 10 years of consistent stakeholder participation in water use and allocation decisions, the interaction between COGERH and Users' Commissions may have created a solid stewardship movement for the preservation of river basin resources that hopefully cannot be reversed.

Yet the model is not without problems. Support for its implementation within the state government has been mixed. While some sectors stand by it, others resist what they perceive is relinquishing too much control over water resources to users (Lemos & de Oliveira, 2004). Moreover, users have also resisted the implementation of the water permit and charging systems. Issues of representation and of equity as well as the enforcement of Commissions' decisions need also to be carefully examined to assess the model's broader effects on the democratization of policy making in the state.[7] Finally, the environmental implications of this model of water allocation and decision making are unclear at this point, especially with regard to the possibility of over-use of water resources. However, even if more research is needed, what has been accomplished so far represents a positive step towards the integration of societal input into water resources management in the state.

The broader impact of interactions such as the one between COGERH and the User's Commission in Ceará is twofold. In terms of policy outcome, it provides a viable model for water management and use in Brazil and other developing countries that avoids both the pitfalls of command and control tools and the shortcomings of market-led privatization schemes. In terms of policy process, it has the potential to democratize policy making positively by pressing forward decentralization, accountability and meaningful public participation in decision making. Understanding the conditions that facilitate public participation is important in terms of its immediate effect on water reform in Brazil and to inform other policy processes and different kinds of public-private partnerships. The Ceará case can offer valuable lessons relevant not only to Brazil but to other less developed countries as well.

The following sections discuss policy making in Brazil in light of the transition to democracy. The second section focuses on Ceará's physical characteristics and the evolution of water management in the state. The third section briefly describes state–society interactive models. The fourth section analyses the Lower Jaguaribe/Banabuiú and Curú river basins Users' Commissions. The paper concludes with some findings of broader relevance.

Ceará, Northeast Brazil

Northeast Brazil, known as the Northeast, is formed by nine states. Most of its territory is located within the semi-arid region, where average rainfall usually does not exceed 800 mm a year. Despite recurrent drought and one of the worst patterns of land distribution in the country, rain-fed subsistence agriculture still employs around 40% of the Northeast's poor. Because of the Northeast's harsh physical characteristics and high vulnerability to drought, the only commercially viable agriculture relies heavily on irrigation, which consumes close to 85% of the scarce water resources available in the region (COGERH, 2000).

Among all the Northeast states, Ceará has the highest relative poverty rate (Carvalho *et al.*, 1993). Dry climate has always been considered the major constraint to economic development in Ceará, as well as the principal cause of rural poverty. Despite significant water storage capacity in reservoirs throughout the state, 92% of farm families do not have access to irrigated land and thus depend entirely on annual rainfall (IBGE, 1996). Multi-year drought events that cause much hardship for both natural and human systems are relatively common. The state has no naturally perennial rivers, and rainfall from the four winter months is expected to guarantee supply for the remaining year, and possibly to subsequent years in the event of an extended drought (Lemos & de Oliveira, 2004). Water resources in Ceará are divided among five large river basins (Jaguaribe, Acaraú, Curú, Coastal Rivers and Poti). The Lower Jaguaribe river basin includes three of the largest and strategically most important reservoirs of the state: Orós, Banabuiú and the newly completed Castanhão (6.7 billion m^3 capacity, the largest in the state). In order to meet the ever-increasing demand from the capital city, Fortaleza, the state government has constructed massive water canals to transfer water from the Lower Jaguaribe/Banabuiú to the coastal region. Such policy has intensified conflict within the river basin between groups who want to curb further water use and government sectors concerned with decreasing Fortaleza's vulnerability to drought.[8]

The Brazilian government first became involved in drought policy for the Northeast in the late 19th century and since then, large amounts of federal money have been spent in the implementation of drought-relief policies. Through the years, the drought industry, as the drought-relief public policy making apparatus is known as in Brazil, has been continually plagued by charges of political manipulation, clientelism and corruption. Since the beginning, drought relief was closely associated with water management and the first government plan to respond to drought implemented in the 19th century focused on the construction of reservoirs, dams and the drilling of wells. Despite the many facets shaping vulnerability to drought in Ceará, policy making mostly focused on the increase of water resources, either through the construction of massive waterworks or attempts to 'make' rain through cloud seeding.[9] Yet by prioritizing 'technical fixes' over much needed structural reform—such as land reform and wealth redistribution—successive governments in the Northeast states managed to attract large amounts of resources to the region without shaking the power base of local elites (Lemos, 2003). Thus in the past 100 years, close to 7000 reservoirs, large and small, have been built in the state, many on private properties (Garjulli *et al.*, 2002). Against this background, it is hardly surprising that, traditionally, water policy has been highly centralized and politicized in the state and that any attempt to reform the system has been met with resistance.

Yet for the past three state administrations drought-related planning has undergone dramatic change (Lemos, 2003; Tendler, 1997; Carvalho *et al.*, 1993). In the wake of Brazil's transition to democracy, a new cadre of policy makers and government officials emerged in Ceará who have been responsible for the broadest political reform in the Northeast. Rather than short-term emergency actions mostly benefiting large landowners, the state government decided to focus on long-term projects associated with communities. A wide range of new programmes emphasized rural development and alleviation of poverty through the implementation of agrarian reform, the creation of irrigated zones, the development of hydrographic microbasins, rational water management, development of micro- and small businesses in the interior, education, basic rural health and sanitation, agro-industry, rural extension, creation of food security programmes, and community development (Magalhães *et al.*, 1991, p. 33). The new programmes encouraged more community involvement in the decision-making process. However, many of these programmes failed or performed below expectations which means that, despite progress, vulnerability of poor populations in Ceará to drought still remains high.

Policy Making across the State/Society Divide

For most of Brazil's history, public policy making has been characterized by limited public access and the predominance of intermediate control channels, notably the techno-bureaucracy, which ultimately functioned as a filter to the issues which might, or might not, get on the governmental agenda (Schmitter, 1971). In this context, studies of public policy making in Brazil have often focused on the relationship between an exceedingly patrimonialist state and a disenfranchised civil society whose interests would only be addressed to the extent that they coincided with the interests of the state.

In order to expand its control over public policy making, the modern Brazilian state established a reserve of technical competence which, allied with traditional political elites, helped to maintain the state's legitimacy (Nunes & Geddes, 1987; Camargo & Diniz, 1989). One consequence of the creation of these 'islands of competence' was the insulation of certain areas and issues from the reach of interests formed outside the realm of the state.[10] Furthermore, the idea of islands of competence created the myth of an infallible technocracy within the government who always knew what was best. This has been very much the case in water and reservoir management in Brazil. The focus on technical fixes has dominated water policy making, especially during the authoritarian governments between 1964 and 1985. Yet, for all its control, the Brazilian state has been unable to address many of the water-related problems plaguing both urban and rural areas.

After the transition to democracy in 1985, two simultaneous trends in Brazil contributed to challenge the dominance of technocracy over the public policy making process. On the one hand, the state's inability to meet the growing demand of basic services revealed the weakness of Brazil's insulated policy-making model. On the other hand, the emergence of independent societal groups that were not shy to demand their rights resulted in growing pressure for policy making reform. The return to free, direct elections added more pressure on different sectors of the state to become more responsive and accountable to societal groups. In the wake of Brazil's latest constitution enacted in the late 1980s, profound changes in natural resources management were introduced in the governmental policy agenda, not the least of which was the need for equitable distribution, rational use and sustainable planning of water resources.

The failure of Brazil's technocratic model of decision making also fostered among policy makers the recognition of the advantages of forging mutually advantageous relationships with groups outside the state, including in environmental-related fields (Watson, 1995; Lemos & Looye, 2003). Thus, in the view of many policy makers, the partnership with independent societal actors would both increase their clout vis-à-vis other sectors of the state machine opposing their policies (especially development-minded sectors) as well as improve policy outcome. In other words, policy makers expected that stakeholder involvement would result in better-designed policies and increasing levels of stewardship from clients who now would have a much higher stake in the policy making process.

Meanwhile, the growing institutionalization of independent social movements also fostered the emergence of numerous groups of stakeholders who were not averse to interacting with the state to achieve their policy goals. Giugni & Passy (1998) attribute the growing institutionalization of social movements to the inability of the modern state to deal with social complexity on its own. To face such complexity, states embark on a process of co-regulation in which organized groups of civil society play an important role both by lending the state legitimacy and by providing specific knowledge the state does not possess. This interaction between state and society suggests the concept of 'conflictual co-operation' to describe the ambivalent relationship between social movements and state institutions. In this relationship, co-operation is defined as "a relationship between two parties based on an agreement over the ends of a given action and involving an active collaboration aimed at reaching such ends" (Giugni & Passy, 1998, p. 84).

Moreover, all over Latin America, the increasing redefinition of political regimes (from authoritarian to democratic), economic systems (from state-led to neoliberal) and the reconstruction of popular participation have redefined popular representation, forging new structures of relations between state and societal actors. Chalmers *et al.* (1997, p. 545) call these structures 'associative networks', which "link state and societal actors—sometimes including popular ones—through interpersonal, media, and/or interorganizational ties. Multiple networks process and reshape contending political claims through relatively open-ended and problem-focused interactions". The emergence of associative networks is related to a series of factors which influence state-society interactions. Among such factors are: (a) decentralization through the creation of multiple centers of decision-making; (b) the impact of new sources of communication and knowledge acquisition; (c) the emergence of new governance paradigms which advocate greater social and popular involvement in decision-making as a means to increase competitiveness and cost-effectiveness of public administration; and finally, (d) "political learning from elites and popular actors induced by the failure of established strategies and institutions to respond to changing political and economic realities" (Chalmers *et al.*, 1997, p. 555).

Yet, such interactions between state and society are neither static nor unique. The literature on public-private partnerships (PPPs) offers plenty of examples of the broad spectrum of such relationships. Indeed, one of the main organizing principles of PPPs is precisely their flexibility to adapt to local needs and resources. Although, traditionally, PPPs were mostly thought of as alternatives to 'big' and inefficient government, increasingly such partnerships are understood as a viable alternative both to outright privatization as well as to public provision of public goods (Lemos *et al.*, 2002). This is particularly true in cases where there is little or no willingness from the for-profit private sector to engage in such relationships or where the implications of outright privatization might be too costly, politically and environmentally.

In this case, the design of institutional arrangements for water management that simultaneously allow for the rational use of resources and promote positive values in public participation is highly desirable. One way to assess these partnerships is to evaluate them in terms of their 'synergy', that is, how embedded and mutually beneficial they are. Peter Evans argues that state-society synergy, "... may be based on 'embeddedness'—that is, on ties that connect citizens and public officials across the public-private divide" (Evans, 1996, p. 1120). Synergistic state-society interactions can significantly support the formation of social capital, avoid traditional practices of exclusionary politics, encourage accountability and improve decision making processes.

Although factors affecting state-society synergy may vary widely, Evans proposes a few critical variables that influence private and societal actors' interactions. In principle, strong, less corrupt and clientelist bureaucracies and more competitive political environments should be more conducive to state-society synergy. Similarly, countries where the existing institutional arrangements encourage community mobilization, transparency, and accountability should fare better on building environments conducive to synergy. Third, there must be complementarity between the private and public sectors to make it worthwhile for groups on both sides to engage in a relationship (Evans, 1996). Finally, Evans identifies the previous endowment of social capital as another factor positively shaping state-society synergy. Next, the water reform in Ceará is analysed and it is suggested that its model of state-society interaction can be expanded to other similar policy environments in Brazil and other less developed countries.

Users' Commissions, COGERH and State/Society Synergy

As mentioned above, water management in Ceará has been a mostly centralized and technocratic affair. However, the enactment of state Law 11.996, which defines the state policy for water resources, changed water management and use in the state. This new regulatory framework has as organizing principles the decentralization, integration, and participation of users in the process of management of water resources, while creating several levels of water management, including river-basin-level Users' Commissions, River Basin Committees, and a state-level Water Resources Council. The law also defines the river basin as the planning unit of action, spells out the instruments of allocation of water rights and fees for the use of water resources, and regulates further construction of infrastructure in the river basin (Garjulli *et al.*, 2001).

To date, the Lower Jaguaribe/Banabuiú River Basin Users' Commission is the most advanced in Ceará, and possibly in Brazil, in terms of organization and user participation. The project began with the creation of COGERH in 1993. However, the implementation of such an encompassing programme in an environment of fierce competition and unequal power distribution was not an easy task. As the authors have argued elsewhere (Lemos &

de Oliveira, 2004), the mutually beneficial relationship between COGERH and local organized groups—especially commercial agriculture, irrigated farmers associations and co-operatives and rural labour unions—was instrumental in the implementation of participatory schemes within Ceará's water reform.[11]

Programme implementation started in 1994 when a team from COGERH's Department for the Organization of Users (DOU) visited the river basins most affected by the 1991–93 drought. Their organizing principle was to employ a multidisciplinary team of technocrats, locally know as *técnicos* (sociologists, geographers, agronomists and engineers), to approach the problem of river basin management from an integrated and participatory framework. Implementation started at the Lower Jaguaribe-Banabuiú valleys, which faced serious water shortage and potential conflict among users as a result of the low levels of their main reservoirs (Alvarez da Silva *et al.*, 1995).

Since the beginning, it was the team's goal to act not as organizers but as facilitators, supporting and stimulating the involvement of groups already organized in water management. Moreover, the team was convinced that in order for the Users' Commission to work, it had to function independently from COGERH. After realizing that they knew very little about local organizations and were unprepared to mediate water allocation among users, COGERH *técnicos* decided to organize an open meeting with users as a first step for mapping out the different kinds of stakeholders in the river basin. COGERH organized the first Seminar of Users of the Waters of the Lower Jaguaribe and Banabuiú river basin. Because they had little experience in the region and very little time, COGERH *técnicos* widely advertised the meeting through radio, newspapers, fliers and direct mail to reach the broadest number of water users possible. Finally, 180 users, representing 63 organizations, attended the meeting. Its main outcome was the creation of an operational plan to manage the river basin. The assembly also elected a 26-member Committee of Representatives to oversee the implementation of the plan. In the second half of 1994, the team of *técnicos* began charting the institutional map affecting water management in the 19 *municípios*[12] within the Lower Jaguaribe/Banabuiú river basin.

Meanwhile, the group met monthly with the Committee of Representatives to discuss the opportunities and constraints for the implementation of the plan and operationalization of effective participatory management for the following year. Most of these meetings were carried out by *municípios* with the intention of developing local capacity and of encouraging participation taking into consideration the local reality of the users. The meetings were organized and led by local representatives from the Users' Commission with the goal to create an oversight organization but also to promote the idea of independence between the Users' Commission and COGERH (Alvarez da Silva *et al.*, 1995). These local meetings provided both for the emergence of new leadership and the expansion of the action arena. One consequence was that when the time came to elect new members for the Users' Commissions, the number of participants in the election process was substantially expanded. As a result, many believe that the new Commission was more representative geographically and sectorally. Another outcome of the local meetings was the preparation of one report for each *município* detailing a wide range of water related issues, including different stakeholders, uses, infrastructure problems, etc. The goal of these reports was to inform COGERH and the Users' Commissions' decision-making process about water allocation and use. In addition, local stakeholders suggested immediate action in three areas of water management: (1) the need to define the operationalization of the several different water uses in the valley and how they affected each other; (2) the need to recuperate and conserve the river basin environmentally using both education and participatory approaches; and (3) the need to train and form a cadre of specialized stakeholders in the areas of environmental conservation and monitoring, river basin management, and rational and sustainable water use to support the Commission's work.

Moreover, COGERH also worked at the state level to publicize the role of the Users' Commission and garner support for the plan from other interests within the state apparatus. The following year, the Lower Jaguaribe/Banabuiú Commission was increased to 53 members to include representatives from all the *municípios* in the river basin. At the same time, Commission members organized local meetings to publicize their work, to attract more participation, and to reinforce their autonomous character vis-à-vis COGERH and its *técnicos*.

While the process of creation and organization of the Curú River Basin Users' Commission was in many ways similar to that of the Lower Jaguaribe/Banabuiú river basin, because of its lower level of stakeholder organization, it required a concerted effort to inform and expand awareness of local stakeholders. Hence, early on, COGERH's DOU's *técnicos* realized that the organization of the Curú river basin would require a more aggressive approach in terms of mobilization, training and education of stakeholders. One particularly challenging constraint was the predominance of traditional agriculture practices which fail to take full advantage of the river basin's irrigated areas and where traditional, low aggregated value crops such as corn and beans and inefficient irrigation techniques predominate. Low productivity and lack of efficient organization for commercialization—such as weak co-operative and farmers associations—resulted in very low living conditions and low levels of mobilization in the Curú river basin when compared with other irrigated valleys in the state (de Oliveira *et al.*, 2001).

From an institutional point of view, the region is closely controlled by the National Department of Public Works and Drought Relief (Departamento Nacional de Obras e Combate às Secas (DNOCS)) an agency in charge of the management of federal reservoirs in the Northeast. Before 1997, DNOCS exerted strict control over water allocation with virtually no participation either from other municipal or state organizations as well as users. The lack of dialogue among the several stakeholders not only generated conflicts but also encouraged the endurance of authoritarian practices such as paternalism, lack of accountability and clientelism (de Oliveira *et al.*, 2001).

Rather than stimulating productivity and acting as a catalyst for social mobilization, the Curú river basin's proximity to Fortaleza worked as one more constraint. Differently from the Lower Jaguaribe/Banabuiú river basin, where the level of mobilization was higher, in the Curú river basin can be found "weak social relations in which the ties of kinship and clientelism prevail over notions of citizenship rights. Apparently, the cultural identity of these *municípios*, which should work as a catalyst for mobilization, is still incipient. In these *municípios*, we did not find a network of solidarity that could function as a mediating instance of the interests and demands of the collective" (de Oliveira *et al.*, 2001, p. 2, authors' translation). Therefore, beyond the organizational efforts employed in the Lower Jaguaribe/Banabuiú river basin, in the Curú it was necessary to lay the foundation more carefully, investing heavily in public awareness and training at the same time avoiding traditional patterns of clientelism and paternalism existing in the region. As a consequence, the process in the Curú was slower and more challenging despite the experience already accumulated by COGERH técnicos. The creation of the Users'

Commission required careful planning and intense work in the field to support mobilization and create a minimum of an institutional and organizational basis that would work as the launching pad for an effective and independent User Commission.

However, the Curú's Users' Commission was short lived. Despite some resistance from water users, following its installation, the River Basin Committee opted to replace the Commission with an Operations Commission within the realm of the Committee. The Operations Commission still had the mandate to negotiate water use and allocation, but differently from the Users' Commission, its membership was limited to River Basin Committee members. Hence, although other users could still participate in the negotiation meetings, they could not vote on the final water allocation.

Despite DOU's early success in stimulating user participation, the decision to create Users' Commissions independent from the state was not without its critics. Many sectors of the state government were profoundly distrustful of the ability of users to make decisions. These sectors were also uncomfortable with relinquishing their control over the decision making process. Another point of contention was whether to establish limits to the kind of decisions Users' Commissions could make. What if Users' Commissions make the 'wrong' decision?[13] For this reason, many technocrats in the state government—and within other sectors of COGERH as well—resisted DOU's efforts to further institutionalize the role of Users' Commissions within Ceará's water management system. This position seems to have prevailed after the election of a new state administration in 2002 that virtually halted the creation of any new Users' Commissions in the state. Recently, however, there have been indications that the 'Users' Commission model' is gaining support again, especially as a result from a new effort by DNOCS to reform its own reservoir management system in other states in the Northeast (Lemos & de Oliveira, 2004). In consequence, the agency is actively organizing Users' Commission in other valleys in the region.

Despite the setback in Ceará, for the past 10 years, the Lower Jaguaribe/Banabuiú River Basin Users' Commission has met regularly several times in the year and more frequently in the months right before the dry season (May/June). Although many aspects of water management are included in the agenda, the main goal of the Users' Commission is to debate and deliberate over the allocation of water stored in the reservoirs for the following year. In order to decide how much water to discharge from the reservoirs and its distribution among several users, the Commission starts to prepare early in the season. Among the subsidies for its decision, a key tool is a simulation prepared by COGERH *técnicos* which is designed to model reservoir discharge and re-charge, as well as the net quantity of water available after each rainy season. These simulations consider a series of parameters and inform users of the several possibilities for discharge given current reservoirs' levels and the prognosis for re-charge the following year.

In 2000, for example, after a year of low rainfall, reservoirs were relatively low (Lower Jaguaribe 51% and Banabuiú 16.4%). Early on, it became clear that the Users' Commission would have to make some tough decisions if it wanted to supply the demand of regular users for the following year and guarantee minimum levels to supply human demand for subsequent years. The Users' Commission met in the city of Limoeiro in June 2000 with the participation of *técnicos* from COGERH and representatives of the main users in the river basin. After an opening section where some technical aspects were detailed—including the simulation of seven different scenarios of discharge and re-charge of the reservoirs—the Commission was divided into groups to discuss the various aspects of the allocation plan. Because the reservoir capacity of the two rivers was so different, the Commission decided to form two subgroups to work separately and independently. The first group, comprised of the users of the Banabuiú River Valley, was excused from having to contribute to the supply of water both to the Lower Jaguaribe Valley as well as the city of Fortaleza and worked on the allocation of resources only for local use. The second group, formed by users from the Lower Jaguaribe Valley, would decide about water allocation among users in whole Jaguaribe valley as well as water supply for the canals serving Fortaleza. Each group then presented the results of its discussion to the assembly, which voted on several alternatives. The Commission decided by consensus that a relatively conservative water volume (average $19 \, m^3/s$ in the Orós and $7 \, m^3/s$ in the Banabuiú) was to be distributed over the next seven months, with allocation levels declining incrementally each month.

In addition, the Commission debated and made recommendations on several other issues, including suggestions for infra-structure improvement, the establishment of more stringent monitoring of users' compliance, and the implementation of education and training projects to improve the sustainability of the water system (COGERH, 2001). This decision meant that some users would have to reduce their water consumption voluntarily, by either decreasing irrigation levels or by cutting back activities (e.g. reducing area planted or having only one crop cycle instead of two per season). Such decisions surprised many sceptical government officials and reinforced the resolve of DOU's *técnicos* to push for further institutionalization of the role of the Commissions within the state's water management system (Lemos & de Oliveira, 2004). Hence, although it may be too early to assess how the decisions of Ceará's new water management institutional apparatus ultimately will affect both policy process (i.e. democratization of decision making) and outcome (i.e. better water management), it seems that many aspects of Evans' state-society synergy proposition can be identified in the interaction between river basin stakeholders and public officials in Ceará.

First, in the case analysed here, there was surely complementarity between COGERH's *técnicos* and the stakeholders at the river basin level. Both groups not only shared goals and responsibility—better water use and allocation, administration of conflict, system sustainability and compliance with regulation—but their partnership contributed to their mutual strengthening vis-à-vis political opposition to their goals. Although it is difficult to establish a clear causal relationship in cases such as these, many at COGERH believe that the interaction with stakeholders was a critical factor enabling water reform implementation. Indeed, many attribute the longevity and success—and hopefully the irreversibility of the new model of management—to the stewardship and commitment of users who after experimenting with participation are unlikely to let it go easily. In addition, essential aspects of water reform success in Ceará depend on the complementarity of roles played by COGERH and the Users' Commissions. Thus by providing users with state-of-the-art information on reservoir management, COGERH enables users to make the best possible decisions within uncertain conditions. Additionally, both groups play important roles in monitoring use and compliance.

Second, there was a strong technocracy committed to implementing water reform. In this case, the team of state *técnicos* appeared to be highly motivated to implement water management reform and especially committed to the most innovative aspects of the new regulatory framework, namely, decentralization, public participation and environmental sustainability. They tirelessly and meticulously planned and implemented their actions while at the same time building up the strength of the Users' Commissions. They were able to garner state support (both in terms of financial and human resources) that enabled policy implementation. COGERH's reputation and clout within the state government also critically contributed to the implementation success. The agency is widely recognized as technically competent and relatively insulated from the most deleterious aspects of backwards politics in the state. In this sense, COGERH's ability to insulate itself as a decision maker is similar to other environmental policy-making agencies whose technical character might render them less vulnerable to traditionally clientelist practices (Lemos, 1998).[14]

Third, political competitiveness and the question of 'rules of the game' seem also to play a significant role in the case reviewed here. Regarding the 'rules of the game', the liberalization of Brazil's political regime and the elimination of most of the repressive barriers that might have hampered free social organization certainly affected the ability of groups outside the state to mobilize and organize. The fact that the last three Ceará reformist governments supported COGERH and other similar participatory policy-making processes (Tendler, 1997; Lemos, 2003) was critical for *técnicos* to be able to keep opposition from local politics and elites at bay. Yet, DOU's inability to push for the institutionalization of the Users' Commissions suggests that there might be a limit for institutional support for public participation (Lemos & de Oliveira, 2004). Moreover, as far as political competitiveness is concerned, public sector workers involved in the projects appeared more motivated to attain their own technocratic objectives than catering to any political faction. Although this is not to say that organizations and individuals were not interested in accruing benefits from project implementation, electoral politics seem to have been a secondary consideration in the process.

Finally, the evidence on endowment of social capital is less straightforward. While it seems that DOU *técnicos* were able to develop solid relationships of trust with most of the users, there is no sound evidence of how critical social capital among stakeholders was a factor shaping the decision-making arena. Information from the river basins studied here suggests that social capital may have been considerably more developed in the Lower Jaguaribe/Banabuiú river basin than in the Curú river basin. Thus while in the Lower Jaguaribe/Banabuiú COGERH *técnicos* found a strong level of previous mobilization and organization, in the Curú it was necessary to lay the foundations that made the organization of Users' Commissions effective. In addition, the fact that COGERH succeeded in supporting a wide array of organizations at the level of both river basins indicates that the agency may have been able to build social capital through its commitment to popular participation in an informed and independent basis. Thus, it seems that the Ceará case reinforces Evans hopeful view that social capital is 'buildable' even in environments where state/society relations had been historically shaped by authoritarianism and clientelism. However, until more in-depth research to assess the level of social capital prior to and after the implementation of the Ceará model is carried out, much of what can be said about its role will be speculation.

Conclusions

Despite widespread recognition of PPPs as innovative and often efficient tools to address the gap between service demand and delivery in less developed countries, one important aspect of this model, public participation, remains elusive. Since public consultation and participation may critically affect the ability of PPPs to work more or less successfully, understanding conditions that influence social actors' ability to work across the public-private divide is essential. This study examines one case of successful state-society partnership formed through the implementation of water management reform in Ceará, Brazil.

By closely examining the formation of Users' Commissions in two of Ceará's most important river basins, it is found that the creation of state-society synergy is an important component of public-private partnerships. Synergy strengthens public-private partnerships through mutually supportive interactions based on embeddedness, that is, ties that connect citizens and public officials across the state-society divide. In Ceará, it is found that such ties are constructed by the systematic organization of stakeholders in a way that at the same time builds their capacity for decision making and reinforces their independence from state organizations. Synergy also improves the likelihood that meaningful public participation will take place both increasing policy legitimacy and stimulating positive values such as transparency, accountability, and equity. It is also found that the four conditions conducive to synergy suggested by Evans' model—that is, a strong and committed bureaucracy, political competition, availability of social capital and complementarity—are either present or may have been successfully built. Moreover, evidence from the Ceará case suggests that another aspect of synergy, less stressed in Evans' framework but critical for COGERH's ability to implement public participation, was DOU's resolve to encourage mobilization and formal organization of users independently from the agency. However, such independence means that COGERH will have to live with Users' Commissions' decisions even when these do not agree with the *técnicos*' best judgement. This can be especially complex in cases where users' decisions for bulk water use might compromise the sustainability of the whole river basin or seriously compromise water availability to all users.

Despite the risk, water users' organization in Ceará holds many promises for water management in Brazil and other parts of the world where water scarcity is critical. It is thought that the model of state-society engagement described in this study can provide a good example for other public-private partnerships. By describing the institutional arrangements shaping meaningful public participation in river basin decision making, this study can critically contribute to the design and implementation of river basin decision models across the public-private divide.

Notes

1. National Council for Public Private Partnerships (2002).
2. This research is partially funded by the National Science Foundation (NSF) and the National Oceanographic and Atmospheric Administration (NOAA). It is also associated with the Watermark Project Project (Projeto Marca D'agua), a broad research initiative to compare water reform across river basins and across time. For more on the Watermark Project, see www.marcadagua.org.br.
3. The Users' Commissions were designed as precursors to the more 'official' River Basins Committees (RBC) which were included in the new state Water Law. Since the early 2000s several RBCs have been or are in the process of being created in Ceará. However, at least in one case, that of the Lower Jaguaribe/Banabuiú valleys, the creation of a RBC did not invalidate the role of the Users' Commission, which continues to operate side-by-side and in conjunction with the Committee.
4. The proportion of representation can vary. For example, in the Lower Jaguaribe the breakdown is 60% users and society and 40% for municipal, state and federal sectors; in the Curú river basin, the breakdown was 50% users and society and 50% municipal, state and federal sectors (Garjulli *et al.*, 2002).

5. In 1997, the federal government enacted Law 9.433 also known as 'Water Law' (*Lei das Águas*) that instituted the National Policy for Water Resources and at the same time created the National System for the Management of Water Resources.
6. The Ceará model was also critically shaped by other outside actors, especially the World Bank. In the early 1990s, as part of the negotiation for a loan to improve the state's water related infrastructure, the Bank imposed as one of the conditions for the loan the creation of a formal water agency to implement water reform. In addition, outside consultants hired to advise on the creation of the water agency suggested that it include social scientists entrusted with the task to design and implement a strong community participatory framework for water management. For more on the creation of COGERH and the role of the World Bank, see Kemper & Olson (2001).
7. Many of these issues are currently being examined in the sphere of the Watermark Project (Projeto Marca D'agua).
8. In 1993, after a three-year drought, the city was threatened with extreme water scarcity. The government implemented a water rationing programme and sped up the construction of a water canal transferring water from the Jaguaribe river basin to Fortaleza.
9. From the late 1950s to early 1990s Ceará invested substantially both in the science as well as in empirical experiments of cloud seeding. At one time, the state had three airplanes dedicated to the task and despite virtually no evidence of success, cloud seeding became an important policy tool because it conveyed the idea of 'action' from the government. In this sense, the sound of the airplanes in the sky became more important than the rains they were supposed to be triggering. For more on cloud seeding and the predominance of technical solutions to water scarcity, see Lemos (2003).
10. In practice, this meant that certain organizations would operate competently almost *despite* the fact they were part of the government. They would be protected from common clientelistic practices such as nepotism, corruption, party spoils, etc. The most notorious and controversial 'reserve of competence' was the apparatus created to design and implement economic policy in Brazil during the military regime. For more on the economic planning institutional apparatus see Dreifuss (1981).
11. However, the predominance of irrigated agriculture and its organized groups has skewed the dispute for water allocation in their favor to the detriment of other uses such as fisheries and rain-fed agriculture (Alvarez da Silva *et al.*, 1995).
12. A *município* is a political subdivision, roughly corresponding to a US county.
13. In at least one documented occasion, some members of the Users' Commission resisted attempts from state officials to curb water use by establishing a programme that at the same time imposed water fees to most users and compensated others for lost yields due to limited availability of water for irrigation. The programme, called Águas do Vale, was designed in response to the drought-induced water crisis of 2001. For more on Águas do Vale see http://www.srh.ce.gov.br/
14. However, the recent election of a new state government may threaten the stability of the current institutional design of river basin management in Ceará. For more details, see Lemos & de Oliveira (2004).

References

Agência Nacional de Água (2002) *A ANA e sua Missão: ser guardiã dos rios, Agência Nacional de Água* (Brasilia, Brazil: Agência Nacional de Água). Available at http://www.ana.gov.br/folder/index.htm.

Alvarez da Silva, U.P., de Oliveira, J.L. & Bezerra, H.E. (1995) A Experiência de Gerenciamento Participativo na Bacia Hidrográfica do Jaguaribe—Ceará, Brasil. Paper presented at the III Simpósio de Recursos Hídricos do Nordeste, Salvador, BA.

Ballestero, A. (2004) Institutional adaptation and water reform in Ceará. Unpublished Master Thesis. School of Natural Resources and Environment, University of Michigan, Ann Arbor, MI.

Bureau for Development and Policy (BDP) (2002) *Public Private Partnerships for the Urban Environment* (Pretoria, South Africa: Bureau for Development and Policy (UNDP)) Available at http://www.undp.org/ppp/.

Carvalho, O., Gonçalves Egler, C. A., Mattos, M. M. C. L., Barros, H., de A Moura Fé, J. & Nobre, C. (1993) *Variabilidade Climática e Planejamento da Ação Governamental no Nordeste Semi-Arido—Avaliação da Seca de 1993* (Brasília: Secretaria de Planejamento, Orçamento e Coordenação da Presidência da República-SEPLAN-PR, Instituto Interamericano de Cooperação para a Agricultura (IICA)).

Camargo, A. & Diniz, E. (1989) Dilemas da consolidação democrática no Brasil, in: A. Camargo & E. Diniz (Eds) *Continuidade e Mudança no Brasil da Nova República* (São Paulo: Vértice).

Chalmers, D. A., Martin, S. B. & Piester, K. (1997) Associative networks: new structures of representation for the popular sectors, in: D. Chalmers, C. Vilas, K. Hite, S. Martin, K. Piester & M. Segarra (Eds) *The New Politics of Inequality in Latin America* (New York: Oxford University Press).

COGERH (2000) *VII Seminário de Operação dos Vales do Jaguaribe e Banabuiú* (Fortaleza: Companhia de Gestão dos Recursos Hídricos, Secretaria dos Recursos Hídricos, Governo do Ceará).

COGERH (2001) *Relatório do VIII Seminário de Planejamento e Operação Vales do Jaguaribe e Banabuiú* (Fortaleza: Companhia de Gestão dos Recursos Hídricos, Secretaria dos Recursos Hídricos, Governo do Ceará).

Dreifuss, R. (1981) *1964, A Conquista do Estado: Ação Política e Golpe de Classe* (Petropolis: Ed. Vozes).

Evans, P. (1996) Government action, social capital and development: reviewing the evidence on synergy, *World Development*, 24(6), pp. 1119–1132.

Garjulli, R., Magalhães, B. S., Teixeira, V. M. R. S., Silva, F. C. B. & Alves, R. F. F. (2001) *Oficina Temática: Gestão Participativa Dos Recursos Hídricos* (Aracaju: PROÁGUA/ANA).

Garjulli, R., de Oliveira, J.L.F., da Cunha, M.A.L., de Souza, E.R. (2002) A Bacia do Rio Jaguaribe, Ceará (Relatório de Pesquisa, Projeto Marca d'Agua) Available at http://www.marcadagua.org.br

Giugni, M. C. & Passy, F. (1998) Contentious politics in complex societies, in: M. C. Giugni, D. McAdam & C. Tilly (Eds) *From Contention to Democracy* (Lanham: Rowan and Littlefield Publishers, Inc.).

IBGE (1996) *Censo Agropecuário de 1995–1996* (Rio de Janeiro: Instituto Brasileiro de Geografia e Estatística).

Kemper, K. & Olson, D. (2001) Water pricing: the dynamics of institutional change in Mexico and Ceará, in: A. Dinar (Ed.) *The Political Economy of Water Pricing Reforms* (Boulder, CO: Netlibrary).

Lemos, M. C. (1998) The politics of pollution control in Brazil: state actors and social movements cleaning up Cubatão, *World Development*, 26(1), pp. 75–87.

Lemos, M. C. (2003) A tale of two policies: the politics of seasonal climate forecasting in Ceará, Northeast Brazil, *Policy Sciences*, 32(2), pp. 101–123.

Lemos, M. C. & Looye, J. (2003) Looking for sustainability: environmental coalitions across the state/society divide, *Bulletin of Latin American Studies*, 22(3), pp. 350–370.

Lemos, M. C. & de Oliveira, J. L. F. (2004) Can water reform survive politics? Institutional change and river basin management in Ceará, Northeast Brazil, *World Development*, 32(12), pp. 2121–2137.

Lemos, M. C., Austin, D., Merideth, R. & Varady, R. G. (2002) Public-private partnerships as catalysts for community-based water infrastructure development: The Border Water Works program in Texas and New Mexico colonias, *Environment and Planning C: Government and Policy*, 20(2), pp. 281–295.

Magalhães, A. R., AbreuVale, J. R., Peixoto, A. B. & de P. Franco Ramos, A. (1991) *Respostas Governamentais as Secas: Experiência de 1987 no Nordeste* (Fortaleza: Imprensa Oficial do Ceará).

Marañón-Pimentel, B. & Wester, P. (2000) Respuestas institucionales para el manejo de los acuíferos en la Cuenca Lerma-Chapala, México, *IWMI, Serie Latinoamericana No. 17* (México: Instituto Internacional del Manejo del Agua).

National Council for Public Private Partnerships (2002) Available at http://ncppp.org/.

Nickson, A. & Vargas, C. (2002) The limitations of water regulation: the failure of the Cochabamba Concession in Bolivia, *Bulletin of Latin America Research*, 21(1), pp. 99–122.

Nunes, E. de O. & Geddes, B. (1987) Dilemmas of state-led modernization in Brazil, in: J. D. Wirth, E. de O. Nunes & T. E. Bogenschild (Eds) *State and Society in Brazil: Continuity and Change* (Boulder, CO: Westview Press).

de Oliveira, J.L.F., Garjulli, R. & Alvarez da Silva, U.P. (2001) Conflitos e estratégias—a implantação do Comitê de Bacia do Rio do Curú (Unpublished document, Fortaleza).

Schmitter, P. (1971) *Interest Conflict and Political Change in Brazil* (Stanford: Stanford University Press).

Tendler, J. (1997) *Good Government in the Tropics* (Baltimore: The Johns Hopkins University Press).

Watson, G. (1995) *Good Sewers Cheap? Agency-Customer Interaction in Low-cost Urban Sanitation in Brazil* (Washington DC: World Bank, Water and Sanitation Division).

Institutional Framework for Water Tariffs in the Buenos Aires Concession, Argentina

LILIAN DEL CASTILLO LABORDE
University of Buenos Aires, Buenos Aires, Argentina

State Reform and Utilities in Buenos Aires City and Greater Buenos Aires

The city of Buenos Aires, with 3 million inhabitants and an extension of $200\,km^2$ was established in 1880 as the Capital of Argentina. It should be noted that between 1880–1994 Buenos Aires did not have the same regime as the other provinces in Argentina, and unlike them, its authorities were appointed by the National Government. It was not until 1994, when the Federal Constitution was amended, that the Capital District gained political autonomy.

The Greater Buenos Aires area, with more than 9 million inhabitants, is an urban sprawl that embraces the outskirts of the city of Buenos Aires. The city of Buenos Aires lies on the right-hand bank of the river Río de la Plata, which at that latitude is approximately 30 km wide. The Río de la Plata supplies this large urban sprawl with 92% of its crude water needs.

With regard to water supply and sanitation, in Argentina the sector used to be centralized in a state-owned company called Obras Sanitarias de la Nación (OSN) (National Water Works). From 1980 onwards, the National Government entrusted this utility to the different provincial authorities, which set up their own corporations, whereas in the city of Buenos Aires and Greater Buenos Aires, the utility was still be provided by OSN.

In March 1991, as a consequence of two traumatic hyper-inflation experiences, the Argentinian government implemented a series of measures that would also influence all the utilities. A fixed exchange rate by which US$1 = $1 was adopted, thereby establishing conversion between national and foreign currencies by means of a federal law. At the same time, the country entered into bilateral agreements for the promotion and protection of

Box 1.

	State Reform Law	
• Statute No. 23 696	17 August 1989	State Reform Act
• Statute No. 23 697	17 August 1989	Economic Emergency Act

foreign investment, as was a common practice among countries at the beginning of the decade in the 1990s. Conditions were then suitable for the privatization of public utilities with chronic deficit financing.

The Market Disembarkation

With regard to the water and sanitation sector, in 1992 the National Government called a tender procurement to privatize OSN by a concession compact.[1] The preliminary stages took place during January 1992, with five business groups submitting their bids. The criteria of selection ranged from the operational to the economic fields. The process of concession first addressed the selection of the operator, the second phase dealt with the technical bids and finally, after short-listing, it focused on the economic proposal. In December 1992, the best bid was selected and on 1 May 1993, the company was transferred to the bid-winning international consortium, headed by Suez-Lyonnaise des Eaux as the operator and main shareholder. The name of the concessionaire was Aguas Argentinas SA (AA) (Argentinian Waters Inc.). The concession granted by the National Government to the consortium in charge of providing both utilities, water and sanitation, ends in May 2023.

Suez, formerly Suez-Lyonnaise des Eaux, is the AA concession operator, with operative, financial and technical accountability according to the concession contract. Suez and AA are differentiated corporations, although Suez should hold no less than 25% of AA capital shares and retain veto power. The concession covered the city of Buenos Aires and the belt of the most densely populated districts in Greater Buenos Aires, an area of 1200 km[2]. In this territory there is a population of approximately 10 million people. At the time of the transfer of the utility from the public to the private operator, 5.7 million people had piped-water coverage and 4.6 million were connected to sewerage. As these figures indicate, it is a large area, perhaps the largest water supply and sanitation concession for a single utility in the world (Rey, 2001, p. 19). However, it did not cover the whole of the districts in the area. Some of these jurisdictions, originally excluded from the tender, later joined the 13 initial districts. An important district in the south of Greater Buenos Aires, Quilmes, was later incorporated, buying water in bulk from the concessionaire. Another district, Morón, retained its own provision of sewerage services and its future expansion. In addition, the utility will continue to receive the bulk sewage of the Berazategui and Florencio Varela districts. At present there are 17 districts incorporated in the concession.[2]

The concession did not cover all the areas served by the former public utility, neither did it embrace all aspects of water management. In the city of Buenos Aires, for example, sewerage and rainfall drainage are one and the same system in the old neighbourhoods where the first works were built between 1870 and 1905.[3] However, sewerage and rainfall are served by different networks, built between 1915 and 1945, in the rest of the city.

Immediately after the concession, the rainfall drainage system was transferred from OSN to the City Hall[4] and, as a consequence, it is not within the current concessionaire's responsibility.

Another feature of utilities is the exclusive character of such services. In fact, since the tender, the Buenos Aires concessionaire has become a monopoly for the whole period of the concession. Therefore, there is no market competition. Although the nature of public services in itself could be regarded as a monopoly, and monopolies are expressly admitted in Argentinian law, there are methods of horizontal competition that could be applied, i.e. dividing the cities into areas and allocating each area to a different utility. The users would then be able to compare the quality of the services and the relation between rates and service. It can be also noted that although it is classed as a private company under Argentinian law, the concessionaire company is not quoted on the local Stock Exchange market. Were it to be so, it would entail the public disclosure of balance sheets every three months as a report requirement. Nevertheless, since it issues corporate bonds, known in Argentina as *Obligaciones negociables*, it is constrained to disclose accountancy books. Therefore, the performance of the concessionaire and the information concerning its revenues are publicly available.

As happened in other privatized public utilities in the country, the corporate consortium has not maintained its initial composition. In recent times the stockholding of the consortium has undergone significant changes that have expanded the participation of international companies at the expense of local ones. Current developments in the country between December 2001 and May 2002 have also confronted the concessionaire with unforeseen challenges.

Institutional Framework for Water and Sanitation Utilities

Given the jurisdictions involved and the assets regime, AA is a national (synonym to federal in Argentina) concession (mainly, Statutes No. 13.577 on National Water Works Regime and No. 23.696 on State Reform), entailing national as well as provincial and municipal responsibilities. The responsible governmental bodies in the Buenos Aires city and Greater Buenos Aires water and sanitation concession are, namely, the National Government, Ministry of Public Works (SOP), Under-Ministry of Water Resources, the Province of Buenos Aires, the Tripartite Entity of Works and Sanitation Services and the city of Buenos Aires. The City Halls of the provincial districts served by the concession have also a legitimate interest in the development of the utility.

The Tripartite Entity of Works and Sanitation Services (ETOSS), the only one set up especially for the Buenos Aires water and sanitation concession, is the agency in charge of both the regulation and monitoring of the utility, and of checking the public and sanitation

Box 2.

Buenos Aires Water and Sanitation Concession
- Decree No. 999/92, 16 August 1992, Legal framework for privatization, Annex I: Regulatory framework for water and sanitation concession, Tariff regime, Sections 43–50
- Decree No. 787/93, 22 April 1993, Approval of Compact between National Government and AA Concessionaire

services supplied by the concessionaire. This special intergovernmental body was established under Annex I, Section III of the State Reform Act, approved on 17 August 1989 (Statute No. 23.696). Later, on 10 February 1992, the agency was set up by the compact entered into by the Ministry of Public Works (SOP), the National Water Works company (OSN), the Province of Buenos Aires and the city of Buenos Aires. The compact was approved by Decree No. 999/1992, of 16 August 1992, and became operational on 20 April 1993. The so-called Regulatory Framework dealing with the objectives and means of the concession were also established under this rule, Decree 999/1992, while Decree 787/1993 approved the appointment of the concession and incorporated the concession contract.

In order to foster the adequate intervention of the parties, the framework of the inter-jurisdictional Regulator was based on the balanced participation on the Board of Directors of the representatives of the three political jurisdictions involved, the National Government, represented by the Ministry of Public Works and the National Water Works utility, the Province of Buenos Aires and the Capital District.

ETOSS is endowed with comprehensive functions and is entrusted with:

- watching the quality of the services supplied by the utility;
- approving the rates and tariffs for the services;
- monitoring whether the utility has complied with the concession contract;
- approving the Users' Regulations;
- demanding reports that will enable the follow-up of the progress of the concession;
- endorsing the improvements and expansion plans;
- hearing the users' claims with regard to billing and service;
- fining the concessionaire in the event of a breach of contract; and
- in the event of a serious failure in the supply which could affect public health, even requesting from the President the placing of the utility in administration.

Bodies such as those of ETOSS have a certain degree of autonomy, derived from the stability of their managerial body, the technical requirements in the process of appointment and the political bearing entrusted in the legal rule which creates the regulatory agency. In most cases they are autonomous, in as much as they are funded by their own resources. Their cash flow could stem from two different sources, namely, it may come from a canon paid by the concessionaire or from a percentage of the tariffs paid by the users. ETOSS belongs to the latter group and receives 2.67% of service invoices, becoming an autonomous entity from both the budgetary and administrative points of view.[5] Notwithstanding the aforesaid (character), the agency functions with that character within the Ministry of Public Works (SOP) of the National Government, which is one of the nine State Secretaries dependent on the President's office, according to Executive Decree 357/2002 (Section VIII) dated 21 February 2002.

In the process of transforming the institutional framework of water supply and wastewater disposal, the state has retained some functions, namely, the outlining of the sectorial policies, and the regulation and monitoring of the concessions. At the same time, the private sector has been entrusted with the tasks concerning the operation and expansion of the utility. As a consequence, water supply and sanitation services are a mixture of private concession and public supervision. The public sector's participation is twofold. In the first stage, it is based on the establishment of an autonomous regulatory

Box 3.

Concession Rules
• Regulation No. 601/99 approved by the Natural Resources and Sustainable Development Ministry (NRS), it established SUMA and patterns • Regulation No. 602/99 approved by the NRS, it established tariffs inputs and periodical revisions mechanism • Decrees No. 1167/97; No. 146/98; No. 1087/98; No. 1369/99; No. 200/00; No. 246/02; and No. 357/02, which deals with different topics of the concession • Decree No. 293/02, set up an ad hoc Commission to deal with 2002 crisis

agency (ETOSS) and in a second and clearly decision-making stage, the National Government (Ministry of Public Works) is the implementing authority which performs the monitoring of the concession compact.

In 1999 the SOP, the national authority empowered to supervise the compact, delegated to its Under-Ministry for Water Resources the monitoring of the Tripartite Entity (Decree No. 20/1999, paragraph 10). Later, the same Under-Secretary was appointed[6] as the national authority for the implementation of the Concession Contract.[7] The same rule (Decree No. 200/2000) authorizes the Under-Secretary to participate, for the Buenos Aires as well as other concessions, in matters related to the tariff regime applied by the utilities as well as in the establishment of the fees and contributions the concessionaires are charged for. The national authority for the concession is also empowered to deal with emergency situations, as has been the case since December 2001.

Users' Participation

In the constitutional amendment of 1994, a chapter was introduced on New Rights and Guarantees. In it, Section 42 establishes the participation of consumer and users' associations in the supervising entities for public utilities. Moreover, procedures for conflict prevention and settlement will be introduced by future regulations within the legal framework of public utilities. At present, users' participation is limited within the Tripartite Entity of Works and Sanitary Services (ETOSS) to the constitution of an advisory body called the Advisory Committee, which has no decision-making power.

Another existing mechanism for users' participation is that of public hearings, which are called when an important disagreement arises between the users and the concessionaire. The purpose of these public consultations is the prevention of conflict as they foster cross-fertilization of proposals, thus bridging the gap between different standpoints and, if possible, sorting out the problem. In June 2000 one public hearing was held to discuss the expansion plans for certain districts of Greater Buenos Aires.

The participation of users should remain within the monitoring and dispute settlement mechanism and not within the decision-making bodies. There are no grounds, for example, for their participation in the executive board of ETOSS, among other important reasons because there is no appropriate mechanism for the election of representatives among eight million users. Several proposals have been brought up at the National Congress which envisage the participation of the users in the regulatory agencies, but hitherto they have not been discussed in Congress.

Box 4.

	Public Participation	
• Argentina	Constitution (1994 Amendment)	Section 42; subsection 2

Economic Policy

Although there is a permanent debate on the accuracy of market policies in the operation of utilities, the trend should be that users should pay for the services they receive, to the benefit of society as a whole. Otherwise, non-users would be subsidizing the operational deficits on the part of utilities. A sound operation should at least aim to put an end to deficits. Even if the user-pays principle seems to be equitable, it appears to be controversial from the point of view of the advocates of a social welfare policy (Matthews *et al.*, 2001). In Argentina, utilities have always been considered a state function within the public domain. The services could be rendered directly by the public administration by means of a state-owned company or indirectly by means of the concession of the public service. In the first half of the 20th century, utilities such as electricity, telephones and railroads were rendered by private concessionaires throughout the country. This was not the case with water and sanitation, which has been a public utility since 1905.

After the Second World War a wave of nationalization took over the utilities, which then became national companies. The government became the owner of the assets and operator of the public services. After almost 50 years of state ownership and administration, the mismanagement of most utilities became a main factor for public sector deficits. Low prices and inefficient operation led to low revenues, which discouraged the extension of the service to new groups of users and the investment in the maintenance of existing works. At this stage, in the last decade of the 20th century, a new demand for private operation of utilities arose and a new wave of concessions began.[8]

The conditions of the Buenos Aires concession have their own features, which will be described later. In the country the main target for the privatization of the water sector, and of other utilities, was that of stopping a chronic and serious deficit. In fact, the water and sanitation public utility (OSN), as well as other national utilities, was unable to cover its costs through water tariffs, although this was not due to low tariffs but to very high costs. In any case, those rates were not based on current water use because there were no water meters and no incentives for their installation and them being read, but on other factors, mainly the size of households (m^2) and other parameters for commercial and industrial users. As a result, water charges were technically scheduled rather as an excise on real estate than as a utility fee.

However surprising for a deficit-ridden operator, in the economic stage of the tender the bidders competed by offering the greatest reduction compared to the tariffs charged by the national utility. To that effect, the concessionaire offered a discount on existing tariffs of 26.9%, by which the multiplier factor of the tariff formula, called K, was 0.731. With regard to the variations in the tariff schedule, which affected the K-factor and the whole tariff structure, the rate charges meant to finance the expansion of infrastructure would fall upon the new users as they joined the service. In fact, it was impossible to invoice different charges, and the fees for infrastructure were drawn up as a fixed rate for current users, in order to finance the pipe extension, whereas for the new users there was a special fee to pay for the connection. An additional charge was levied in order to finance a set of new sewerage works that supplemented the contract of concession.

An issue dealt with in the general legal framework for concessions, the State Reform Act, concerned the profits of concessionaire companies. Although utilities are corporations, their business is tightly linked to the social constraints that are implicit in the operation of public services. Thus, profits should not surpass a reasonable relationship between investments actually made by the concessionaire and the net revenue of the concession.[9]

Concerning the charges that the concessionaire of the utility pays to the National Government or the provincial or municipal administrations, the State Reform Act states that concessions can be either priced or free.[10] The general legal framework for concessions is applicable to all utilities, including water services.[11]

Another feature in the concession which was carried out in the city of Buenos Aires and Greater Buenos Aires was the treatment of real estate. It was established that the National Water Works (OSN) public corporation's real estate and assets, amounting to US$1.2 billion according to some estimates, is to remain national property, without any kind of canon being paid by the concessionaire, who is obliged at the termination of the concession to return the assets to its public owner. Consequently, the utility would not be able to obtain funds by selling state assets, nor could it engage in other forms of enterprise characteristic of the private sector.

Furthermore, if these limitations the priority assigned in the concession contract to expanding the services were added up, it could be concluded that a large investment would be inevitable on the part of the concessionaire from the very beginning of the concession. It should be stressed that the goals for extending the service would require investments of almost US$4 billion throughout the concession with an investment of US$1.084 million required in the first five-year stage.

A further subject is that related to the concessionaire's contributions. The minimum contribution required for the concession contract, which amounted to US$120 million, was invested entirely by the consortium winner of the tender in the second stage of the concession, including 10% of the 'Stock Option Program'. The Program was especially included in the Regulatory Framework for those workers of the public utility who would remain in the private corporation. An early retirement scheme was offered to those not wishing to continue working with the new operator. The initial contribution was enlarged with a supplementary sum of US$30 million, and the entire company capital now amounts to US$150 million.

Box 5.

Economic Policy
- Decree No. 999/92, Legal framework for water and sanitation concession
- Decree No. 787/93, Approval of Compact between *National Government and AA Concessionaire*
- Decree No. 149/97, Concession Compact amendment focused on social interests

Water Price and Water Tariffs

Water Price

Utilities are water suppliers for the customers of water and sanitation services, although they themselves are bulk users of water resources as well. This means that they are water producers, that is to say, an industry that transforms raw water into potable water, delivers drinking water to customers and discharges wastewater. Water resources are their raw material (Perry *et al.*, 2001). To assess the significance of the intake and the discharge into the Río de la Plata river currently managed by AA, it can noted that the production of drinking water amounts to 4.1 million/m³ daily and the pumping out of wastewater is approximately 2.3 million/m³ daily.[12]

Rivers and lakes are public goods within the national or provincial jurisdiction in the country.[13] The Río de la Plata, the source of water for Buenos Aires city, is a national (or federal river) like all other navigable rivers in the country.[14] In the AA concession, there is no charge for the intake of untreated water from the Rio de la Plata, or the abstraction of groundwater in the few cases where it is needed, as well as for the discharge of wastewater in the same river. Neither is there a charge for the use of the city underground for the installation of pipes.[15]

Whether to enter into a free or a charged utility concession is a choice for the Argentinian state, as both options are stated in the legal framework. Additionally, there is no mandatory rule establishing a priority between those options. The inclusion of the condition of 'free of charge' in four clauses of the water concession could be construed as an indication that concessions will not necessarily be presumed to be free of charge for the concessionaire of water resources services from the part of the public authority. On the contrary, the clause is favourable to the conclusion that there could be a charge for abstraction and discharge of water unless otherwise stated.

Water Tariffs

Tariff structure. Tariffs are the core issue of the concession, the pivot that turns the concession into either a profitable business or an unsuccessful venture.

In the concession contract, as well as in Resolutions No. 1103/98 and No. 602/99, it was stated that tariffs should cover operational costs, taxation, part of investment costs and capital earnings. The tariff scheme succeeded in covering those items within the eight-and-a-half years in operation until December 2001. The system used is a 'price cap' to be discussed every five years, six times in the concession period, with annual corrections for costs variations. The charges are based on property quality, taking into account zoning, building and size. There is a basic tariff that is invoiced bi-monthly to users (TBB), whether domestic or non-domestic. At the same time, there are other services such as water supply to ships and bulk water for special purposes, which are not included in the regular billing.

The tariffs resulting from the complex formula for charges is proposed by the concessionaire to ETOSS, where it is analysed. The proposal by the utility is the first step to beginning the discussions of the different charges invoiced to customers. After this process and with the opinion of ETOSS, the proposal is sent to the government, Ministry of Public Works, for further consideration. The tariff is finally approved by the National Government, which is the granting authority.

There are three types of rates for the almost 3 million customers and 8 million users: (1) household; (2) non-household; and (3) unused plots. The basic rate is practically the same for all of them. Moreover, charges are unmeasured in the largest proportion. A particular feature of this concession is the high demand of consumers, with an average consumption of more than 300 l/h/d[16] in the whole area, including a higher demand of 400 to 500 l/h/d in the City of Buenos Aires and a lower demand of 200 l/h/d in the outskirts of the city. The river Río de la Plata is the main source of water supply for the Capital District.

The indicative units for rates are US$0.2 to US$1.78/m³ according to occupancy, for a range from high occupancy, high-use customers to low-use single occupancy, and before currency flotation. These figures include both services, water and sanitation, and additional charges.[17] Unlike other cities, where sewerage charges amount to four or five times the potable water supply charges, in the Buenos Aires utility water bills include water and sanitation services equally. This is a mirror of the actual situation of 8 million water and sanitation users with only 10%, 800 000 people, connected to sewage pre-treatment or treatment works. Regrettably, 90% of sewage is pumped directly into the Río de la Plata river.

In the first five-year period invoicing was considerably lower than originally estimated, not only because of unrecoverable collectibles and users in arrears but also due to the inadequate database. With regard to the water bills payment index, it is in the region of 80% to 85% of total invoicing, which could be considered satisfactory. However, the possibility that the concessionaire can interrupt the service due to unpaid bills is highly controversial. Although such a consequence is foreseen in the concession contract[18] there is a mandatory conciliation procedure between debtors and the concessionaire in order to avoid cutting off water supply. In May 2002, a statute was passed forbidding the interruption of water services to schools and hospitals.

An efficiency factor was also incorporated into the tariff schedule, which should come closer to those of the Model Company. The clause is applicable for both absolute and relative efficiency and is to be applied on a yearly basis. It was agreed that there would be a 0.5% annual reduction in tariffs as a result of this efficiency factor.

VAT in Argentina is a national excise and it is applicable to water and sanitation rates, excluding only the 2.67% charge for ETOSS. It is within the responsibility of AA to collect this tax, which ranges from 21% to 39% of total invoicing (27% on average) according to the different categories of users.

Although VAT is not a charge for the concessionaire but for users, taxation is a significant burden for the utility, and it amounts to almost one-third of the concession turnover.

Since the beginning of 2002 the above-mentioned patterns for tariffs have been greatly distorted and there is currently a request from the concessionaire to set new sustainable conditions, beginning with friendly talks that the legal (Statute No. 24.100 and Resolution NRS No. 601/99) and institutional (ETOSS and Under-Ministry of Water Resources) framework foresees.

Modification of tariffs. From the beginning of the operation on 1 May 1993, there were discrepancies between expectations and facts. The concessionaire estimated the revenues according to the database gathered by the governmental agencies, not the public utility, for the tender procurement. Afterwards it became clear that the database was inaccurate and,

Box 6.

Legal Structure of the Tariffs
- Decree No.1167/97, Approved compared—introduced universal service and environmental charges (SUMA)
- Resolution No. 601/99, Annex IV—patterns and mechanisms for tariffs
- Resolution No. 602/99, Economic and financial balance
- Resolution No. 1111/99, Brought the re-negotiation of the Compact to a close
- Compact-minutes of the ETOSS-AA meeting of 9 January 2001

as a result, there was a miscalculation of the future incomes of the utility. The difference had a negative effect on the revenues of the concession during the first five years, because the invoicing during that period was substantially below the bid offer. Therefore, in 1994, AA requested a revision of rates and, in its statement before the Regulator, estimated a necessary rise in the K factor by 30%, taking into account household invoicing and by 41% considering the total invoicing. At that time ETOSS rejected the tariff modification on contractual grounds, but afterwards the subject was discussed in the two extraordinary revisions and during the re-negotiation period from 1997 to 1999.

AA and ETOSS differed in their interpretation of the terms of the contract with regard to the existence of a guarantee clause. For the concessionaire, it was clear that there is a revenue guarantee in the essence of the concession while for ETOSS the concession is inserted in a market economy with no guarantees where the bidders should undertake the risk assessment. There is no guarantee section in black and white in the contract, but there is a mutual commitment of the parties to make the concession work. A company sliding into default is neither a sound management nor a successful achievement.

In Argentina a number of compacts with privatized utilities have clauses which allow the modification of tariffs according to other countries' indexes. This was the case with the water and sanitation concession, where tariffs could be revised according to the price indexes of the USA (the average of Producer Price Index—Industrial Commodities and Consumer Price Index—Water and Sewerage Maintenance, from the Bureau of Labor Statistics Data of the USA). The revision of tariffs was to take place on an annual basis, to be applicable every year on 1 February. This clause turned out to be rather unfair since tariffs could be increased even if prices in other countries did not increase as much or remained steady. The claims raised by users gave support to people opposing the privatization of utilities.

The institutional mechanism for adjustment of concession conditions and tariffs is to re-open negotiations with the purpose of amending some aspects of the compact. There have been several rounds of negotiations with that aim. The modifications included a set of measures aiming to keep the economic and financial balance of the concession. In order to reconcile the different goals of government and concessionaire, some working objectives were reconsidered. The investments made were accepted and considered valid under the contract, some unfulfilled duties were cancelled, some fines were written-off in exchange for future works that were incorporated into the contract, other charges were shifted onto the tariffs and new environmental targets were recognized.

A performance assessment was to take place every five years, and the first assessment began on 30 April 1998 and was extended (Decree No. 1167/1997 of 7 November 1997) until 31 December 1998. This assessment would provide the data for the round of negotiations that should occur five years, according to the concession contract.

The charge rate of infrastructure (CI) was abandoned in 1995 and replaced by a new charge called 'Universal Service' (SU). An extra charge for the environmental programme (MA) was added (Boyd & Ingberman 2001). In 1997 the re-negotiation also included a change in the schedule: the first five-year period was extended several months, which were taken off the sixth, and last, five-year period of the present concession.

On 9 January 2001, the inappropriateness of some conditions included in the concession compact was recognized and accordingly a new round of negotiations took place between governmental agencies and AA. The main outcome was a rise in tariffs by 3.9% starting in February 2001. Such a rise of 3.9% took place anew in the years 2002 and 2003, with a view to launching an investment plan of US$1.016 million to expand the services network "in order to reduce the existing sanitary vulnerability". These increases of 3.9% are to be deposited in a trusteeship and they will finance the works of expansion.

Under such expansion,[19] 1.5 million inhabitants of 17 districts of Greater Buenos Aires are predicted to gain access to the 13 000 km drinking water and 8000 km of sanitation systems in seven years' time. Approximately 470 000 users will be connected to the potable water network after an investment of US$210 million.

The users that will benefit most from the sewerage system are inhabitants of the south area of the city of Buenos Aires, where the greatest sanitary risks are present. In addition, US$190 million will be allocated to build a wastewater pre-treatment plant in the area of Barracas where a pumping plant is in operation at present.[20]

As previously stated, the concessionaire won the tender procurement by a combined offer of better operational skills, excellent technical performance and low tariffs. After nine years of operation, and despite ongoing inflation in 2002, the tariffs are lower than those of OSN before the concession, as it is shown in Boxes 7 and 8.

Investment schedule.   A constant problem during the period of public operation of the water and sanitation utility, the final utility in the hands of the state, was the low investment in maintenance and new works. In the first five-year period of operation, for instance, the investments made by Aguas Argentinas in the specific area covered by the concession were equivalent to five times the average of the previous public operator five-year period.

Before privatization, the disinvestment negatively affected both the expansion of the service and the network. As an example, at the beginning of the concession, leakage was calculated to be in the region of 40–45% of the total production of water. Currently, due to the low micro-measurement in the system, it is considered to be in the region of 35%, according to estimates.

The former 1949 regulatory framework of OSN (National Water Works)[23] enabled the state company, an agency within the structure of the Ministry of Economy, to resort to

Box 7.

	May 1993	April 2002	%
K Index[21]	0.7310	0.9572	30.9
Consumer Price Index	93.04	118.23	27.01
Producer Price Index	98.00	160.69	64.0

Box 8.

Modification of tariffs
- Decree No. 149/97, Concession Compact amendment focused on social interests
- Decree No. 1167/97, Revision of tariffs for cost variations
- Regulations No.1103/93, No. 601/99; No. 602/99; No. 1199/99 (NRS[22])
- Compact-minutes of the ETOSS-AA meeting of 9 January 2001

subsidies when its own income was insufficient to cover its costs. At that time, the government was responsible for fixing the prices and drawing up the investment policies; consequently, operating deficit financing was funded by the National Treasury (Mairal, 1994).

Since privatization, the regulatory framework for the concession has changed the focus. It is clear that its main purpose is to improve the service and to expand the network. Therefore, expansion has regained strength and the area of coverage has grown.

On 30 October 2000, Aguas Argentinas opened the first stretch of the 15.3 km underground pipe, which will enable the supply of potable water to 1.5 million inhabitants of Greater Buenos Aires, with a total investment of US$120 million.[24]

Those areas that are not covered by the concession are able to use wells made by the inhabitants. These wells are possible due to the availability of groundwater near the surface, although, as it contains arsenic and nitrates, this is low quality water. Under the terms of the concession compact the concessionaire is obliged to gradually block the wells and make a mandatory connection to pipe water. Nonetheless, the utility staff admit how difficult it is to detect all the wells and to comply with this contractual duty. There are similar occurrences with the septic wells in the network, which function as inferior quality substitutes for sewerage.

On the other hand, with regard to investment duties, the utility has incurred certain breaches of the contract which have been punished with fines. For instance, AA invested in areas in which it was not contractually bound, such as improvements in the computer equipment for the commercial management in order to build better records of its users, while it did not invest in areas where it is under obligation.

Part of the problem stemmed from the impossibility of financing the network extension with the rates that were charged to the users for infrastructure. Moreover, there was a clear underestimation of the creditworthiness, of the willingness-to-pay or of both, which resulted in delays in debt collection and in the investment plans.[25]

In December 2001, the foreign financial debt of the company amounted to US$706 million. This was a result of the outsourcing of the agreed investment programme. The main creditors are the Inter-American Development Bank (IADB) (US$80 million, first loan agreement in 1995 and US$215 million, second loan agreement in 1999), the International Financial Corporation (IFC) (US$172.5 million in the First Investment Agreement of 1994 and US$213 million in the Second Investment Agreement of 1996, both partially paid off), the European Investment Bank (EIB) (US$77.4 million) and the ING Barings (US$108 million).

The concessionaire was also authorized to issue bonds until the amount of US$500 million. The first series (Series 01 Bonds) of US$90 million was launched in 1997 and was completely redeemed on 2 January 2001. Despite all this, according to corporate accounts, the incomes of the utility reasonably increased until 31 December 2001.

Box 9.

Investment schedule
- Decree No. 999/92, Legal framework for privatization, *Annex I:* Regulatory framework for water and sanitation concession, Tariff regime, Sections 43–50
- Decree No. 787/93, Approval of the National Government—AA Concession Compact
- Compact-minutes of the ETOSS-AA meeting of 9 January 2001
- AA Balance Accounts by 31 December 2001

The concession as a process. In the field of utilities, many circumstances have to be taken into account. It is not just fees and efficiency, legal regime and sanctions. It is a whole set of feelings with regard to the state's duties and its commitment to the well-being of citizens and to the willingness to reduce the gap between the 'haves' and the 'have-nots'. Does the poor citizen have any expectation of having access to drinking water supply, electric power, communication, mail or hospital care? Could someone that has none of those expectations be considered a citizen? In developed democratic societies, taxation is the proper way to balance social disparity, but this is not the situation in an overwhelming number of states. The answer to the question of who has to pay for what is a highly debatable issue.

In Argentina, a society with historical surges of public and private utilities, private at the beginning of the 20th century, public after the Second World War and privatized again at the end of the 20th century, there is little awareness of the relationship between efficiency and price. The idea that public service is a duty for the state and a right for the people is quite different from the attitude of demanding efficient service and reasonable tariffs. Whereas the former concept hints at uneven service with non-profitable political fees, the latter points to better services and rates determined from the costs of the operator (Dinar, 2000).

It seems that there are different situations in each country providing the elements to adopt one policy or the other. The most important could be the income per capita and the relative size of social strata. It is very difficult to find the most appropriate answer, and it is clear that there is no permanent solution. It is known that planning is a never-ending activity, because new facts constantly come to the fore that need to be taken into consideration. There is planning and there is adjustment, and this is the process that a concession of public service undergoes.

At present, about 7.5 million inhabitants are supplied with potable water and about 5.6 million with sewerage as well. The networks comprise 13 000 km of water pipelines and 8000 km of sewerage (Rey, 2001).

With regard to the economic and financial outcome of the utility, company invoicing increased from the beginning of the concession until 31 December 2001. From US$540 million (plus VAT of 21% for final consumers) invoiced in 1999, it reached US$555 million in 2000 and US$650 million in 2001. The estimate for 2002 is in the region of pesos $690 million, which at the current exchange of US$0.29 = 1 peso amounts to a mere US$200 million.

Annual investment for the period 1994–2003 has been estimated at about US$200 million per year. At the end of the eighth year these targets were reached, with an effective investment for the first five-year period of the concession (May 1993 to December 1998) of US$1169 million.

Box 10.

Investment 1993–98
- Expansion of drinking water and sewerage network   US$527 million
- Renovation and rehabilitation   US$144 million
- Operational improvements   US$348 million
- Financial and other costs   US$150 million

For the period 1999–2003, an investment of about US$1000 to US$1100 million is envisaged. Boxes 11 and 12 show the apportionment.

The Argentinian Crisis (December 2001–May 2002)

In 2001 the economic and political situation in Argentina was under severe stress. At the beginning of that year an important 'yield' from international financial institutions, similar to those granted to other countries, was not enough to reverse three years of recession. In December 2001 the government collapsed and a new President was appointed by the National Congress on 1 January 2002. On 6 January 2002, the Congress passed Statute No. 25.561 which ended the 10-year pattern of fixed currency exchange.

The situation of privatized utilities has become difficult, because the unforeseen events surpassed any contractual forecast, rendering it necessary to revise anew the existing conditions. The new set of negotiations should deal with several issues:

1. To deal with short-term emergencies. There are immediate problems such as the renewal of guarantees from private financial institutions, which amount to US$150 million annually. Due to the risks involved, if the guarantees disappear the contract could be invalidated. The utility is currently in default with these payments, and the situation should be resolved without delay. But there are other equally serious and urgent problems as well.

2. To introduce economic and contractual corrections, in as much as a new balance has to be accorded in order to maintain the concession. The conditions seem to be no longer equitable for the utility in order to keep its qualified staff, the high standard of service, the payments of strategic inputs and of taxes. The company is up-to-date with its taxation payments to May 2002. The situation could be considered a remarkable commitment on the part of the utility to overcome the current economic turmoil.

3. To address the very serious problem of the corporate debt. The company is in a default situation with the payments of the interest of its debt with foreign

Box 11.

Investment 1999–2003
- Expansion of drinking water and sewerage network   US$319 million
- Renovation and refurbishment (rehabilitation)   US$473 million
- Integral Sanitation Plan (PSI)   US$189 million
- Environmental objectives and other investments   US$125 million

Box 12.

Services data	The concession as a process		
	1993	1998	2001
Water supply coverage	65%	79%	–
Sanitation	54%	61%	–
Customers (non-users)	2.036.000	2.576.000	647.800
Customers with measured charges	30.000	–	343.800

financial institutions. It is a difficult situation because payments should be made in foreign currency and the new conditions established by Act 25.561 banned transfers abroad. Moreover, the relationship between revenue and debt has changed dramatically.

4. To manage the sharp devaluation of local currency, which was already more than three times its value in only five months since the end of the fixed parity (US$0.294 = 1 peso) and that could not be transferred to utility charges either. A new tariff regime was drawn up to become effective from 2002, but its enforcement has been postponed. Statute No. 25790 dated 1 October 2003 extended the revision of rates to 31 December 2004. Tariffs should remain unchanged in the meantime, according to the Act dated 11 May 2004 between Government and Aquas Argentinas.

There is an intricate fabric of legal instruments, both national and international, affected by and with effects on the crisis in Argentina. Hopefully, the Argentinian economy will be resurrected again and the concession, as well as other concessions and with updated compacts, will survive.

Notes

1. To establish the terms of the concession a *Technical Commission for the Privatization of OSN* was established in 1991 by Resolution No. 97 of the Ministry of Economy, dated 16 April 1991, published in the Official Gazette on 10 May 1991.
2. The 17 districts are: Vicente López, San Isidro, San Fernando, Tigre, San Martín, Tres de Febrero, Hurlingham, Ituzaingó, Morón, La Matanza, Esteban Echeverría, Ezeiza, Almirante Brown, Lomas de Zamora, Quilmes, Lanús y Avellaneda.
3. The old works covers 3000 ha of the 20 000 ha of the city.
4. Decree 993/93, published in the Official Gazette on 21 May 1993.
5. According to estimates, ETOSS' budget for the year 2001 was in the region of US$10 million with a staff of 110 employees. To make a comparison, this significant figure amounts to one-third of the US$30 million annual budget for the utility's sewerage network expansion.
6. Decree 200/2000, partially amending Decree 20/1999, Section 1, subsection 11 on concession's authority.
7. Approved by Decree 787/1993.
8. In Argentine law, which follows the French Administrative Law model, concession is a contract whereby the government grants to a private party the right to operate a utility whose assets belong to the public domain and should be returned at the end of the concession.
9. Statute No. 23696, Section 57.
10. Statute No. 23696, Section 58, second paragraph.
11. Decree No. 999/92, Section 7.

12. The discharge into the Río de la Plata river is made at four points: (1) The north wastewater plant, discharging approximately 70000 m³/day; (2) The south-west wastewater plant, discharging approximately 160 000 m³/day; (3) The Berazategui plant, pumping approximately 1.9 million/m³/day of untreated wastewater; and (4) The New Dock pumping plant, receiving the unmeasured discharge of the oldest works of the city of Buenos Aires by means of three mains receiving sewage and rainfall jointly.
13. Section 2340, subsections 3rd and 5th of the Civil Code.
14. Argentina, National Constitution, Section 75, subsections 10th and 18th.
15. Decree 999/92, Chapter IV, of the concessionaire, Section 29 on Duties and functions of the concessionaire, subsections (r); (s); (t); and (u).
16. l/h/d: litres per inhabitant per day.
17. The additional charges included infrastructure charges until 1995, when it was abandoned and different charges were included, namely, the expansion charge for marginal areas (SU) and the charge for environmental issues which have not been previously considered (MA). VAT at an average of 28% is applied to SUMA charges. There is also a charge to finance the regulatory agency, ETOSS.
18. Section 52, Decree 999/92.
19. Improvement and Expansion Programme: PMES (Plan de Mejoras y Expansión del Servicio).
20. Article published in the *La Nación* Buenos Aires' newspaper on 10 January 2001.
21. K index: Index applied to sanitation services and to be determined by the Regulator.
22. Ministry of Natural Resources and Environment, at present Ministry of Sustainable Development.
23. Statute No. 13577, Legal Framework for National Water Works, 29 October 1949, *O.G.* 2 November 1949.
24. The news was published in the *La Prensa* Buenos Aires' newspaper that same day.
25. Aguas Argentinas SA made a tender to select law firms for debt collection and since 1996, 45 000 law suits were filed before local courts. The residual OSN is as yet following proceedings for unpaid billings before privatization.

References

Boyd, J. & Ingberman, D. (2001) The vertical extension of environmental liability through chains of ownership, contract and supply, in: A. Heyes (Ed.) *The Law and Economics of the Environment*, pp. 44–70 (Cheltenham: Edward Elgar).

Dinar, A. (Ed.) (2000) *The Political Economy of Water Pricing Reforms* (New York, NY: World Bank and Oxford University Press).

Mairal, H. A. (1994) The Argentine privatisation program: model and legal issues, in: M. E. Brown & G. Ridley (Eds) *Privatisation—Current Issues*, pp. 57–73 (New York, NY: Aspen Publishers).

Matthews, O. P., Brookshire, D. S. & Campana, M. E. (Eds) (2001) *The Economic Value of Water: Results of a Workshop in Caracas, Venezuela*, Albuqurque: Water Resources Program, The University of New Mexico).

Perry, R., Blatter, J. & Ingram, H. (2001) Lessons from the spaces of unbound water for research and governance in a glocalized world, in: J. Blatter & H. Ingram (Eds) *Reflections on Water*, pp. 321–340 (Boston, MA: MIT).

Rey, O. (2001) *El Saneamiento en el Area Metropolitana, Período 1993–2000. Los primeros siete años de Aguas Argentinas* (Buenos Aires: Aguas Argentinas).

Private-sector Participation in the Management of Potable Water in Mexico City, 1992–2002

BORIS MARAÑÓN, in collaboration with THE THIRD WORLD CENTRE FOR WATER MANAGEMENT

The Third World Centre for Water Management, Los Clubes, Atizapan, Estado de Mexico, 52958 Mexico

Introduction

Mexico City's (from hereafter referred to as Federal District) new strategy was launched in 1992 by the government of the Federal District in an effort to promote a profound structural change in the city's water management. The thinking behind it was that water could no longer be regarded as a public right (and, as a result, supplied and subsidized by the state), but as an economic asset, subject to private appropriation. The hydraulic system faced a severe crisis as a result of deterioration of the physical and commercial infrastructure, inefficiency and a charging system based on fixed tariffs. There was also a need to eliminate subsidies that promoted waste while blocking expansion and the improvement of supplies to the poorest neighbourhoods. It was also necessary to develop participation schemes for the private sector in different stages of the production, distribution and sale of water (CADF, 1993, pp. 2–3).

The new strategy had two objectives: guaranteeing the water supply that the city needed for its development on an ecologically sustainable basis, and achieving financial self-sufficiency for the system (CADF, 1994; Ministry of Finances (SF), 1997, p. 20). Both immediate and long-term measures were planned in order to achieve these goals. The immediate measures included an updating of the legal and institutional framework, including a definition of the policy to be followed on buildings, powers to curtail or cut off the supply, and charges for rights of discharge into the drainage system of water extracted from wells (the only previous charge was a one-off payment to the National Water Commission, the CNA). At the same time, control was sought over the 10 000 major users, who provided more than 60% of income (SF, 1997, p. 23).

However, these measures were not sufficient to achieve ecological and financial self-sufficiency. As a result the government had to promote two far-reaching measures that, in the end, proved to be the key to the new water strategy: (a) the introduction of a system that based all charges on metering and (b) a massive rehabilitation programme for the distribution network in an attempt to reduce losses through leakages to between 10 and 15% (CADF, 1994; SF, 1997, p. 25). The government also thought it was necessary to eliminate the overlapping of functions by all the various bodies involved with water, and bring the sector's finances into balance in the medium term. Integration, it was thought, would improve co-ordination between the functions of distribution and charging, as well as providing the necessary incentives by establishing a relation between spending and income. The aim was to create an institution that, either by itself or acting through third parties, would provide the public services of drinking water, drainage and treatment and re-use of wastewater, as well as operating, managing and conserving the hydraulic system (Beristain-Iturbide, 2002). The new body was to be semi-autonomous and incorporate all the necessary functions and faculties, while making as little impact as possible on those of the Department of Hydraulic Construction and Works (DGCOH) and keeping to a minimum any problems that might emerge on the labour front. The main function of the new body, however, was to be the introduction of a billing system based on metering, so as to bring the water and drainage budget into balance as quickly as possible. The private sector was to be invited to take charge of distribution, metering, billing, customer support and maintenance of the secondary network. At the same time, the city treasurer's office and the boroughs were to reach accords on charging and the installation of uptakes, pending the assignation of zones to private operators.

On 14 July 1992, the president of Mexico issued a decree creating the Federal District Water Commission (CADF) as an autonomous administrative body that would take overall charge of the provision of the public services of drinking water, drainage, treatment and re-use of wastewater, and would unify all efforts and measures taken in these areas (CADF, 1993, p. 7). For socio-political reasons (staff cuts and subsequent protests) and financial considerations (a reduction in income), a decision was taken to integrate the various functions one step at a time. It was thought that, in the medium term, the CADF would evolve into a body that would take in all aspects of management of the service. This thinking was justified on the basis that the conversion of rights into tariffs would make it easier to update and administer this aspect of the work, and because greater financial self-sufficiency would be achieved by the linkage of the sector's spending and income (Beristain-Iturbide, 2002, pp. 6–7).

The Contracts: Phasing-in and Fragmentation

From the beginning, private sector participation in the Federal District was unorthodox. On the one hand, the private sector was gradually to take over responsibility for the system, beginning with service contracts. On the other hand, not one company was contracted but four.[1]

According to the CADF, in order to meet the condition of keeping the service under the responsibility of the city government, the use of such traditional schemes of private-sector participation as the sale or concession of infrastructure was considered inappropriate. Instead, private-sector participation was structured through service contracts under which property rights in the infrastructure and control over the introduction of the new system of charges would remain in the hands of the city government (CADF, 1995, p. 21). It was also decided that the application of metering, and improvements to the system, would be carried out with the help of the private sector.

The next step was to decide the extent to which private companies could participate in the sector. This was defined on the basis, on the one hand, of the political decision to maintain control and responsibility for the service in the hands of the city government and, on the other, on the lack of accounting, financial and operational information on the system. In order to keep the service as a responsibility of the city government, the contract that was considered to be most appropriate was of a type similar to the French *affermage*. Under this format, the city government could retain control and responsibility for the system, as well as property rights in the infrastructure and the power to set tariffs. The contract that was finally signed was generic and atypical. The Public Works Law and its regulations and complementary dispositions had to be applied in a way that reflected its *sui generis* nature (SF, 1997, pp. 17–18).

At the same time, in order to phase in the project by stages, a juridical scheme had to be developed that would facilitate progress with the support of a normative contract that covered all stages before the risk contract, and each of these stages had to be covered by specific contracts. Under this scheme, all necessary information could be gathered before proceeding to a contract based on performance incentives. Operation under service contracts or by specific tasks would allow the city to control the pace and the policy for the introduction of new charging systems, therefore reducing the financial uncertainty and the political risk, both for the contractor and for the city government, before the adoption of a risk contract (SF, 1997, pp. 18–19).

Three stages were defined in keeping with the policy of phasing in the new system. In the first stage, contracts were to be issued for the installation of meters and billing systems based on them, as well as the drawing up of registers both of users and of the primary and secondary networks, and the regularization of uptakes. Ideally, this was to be done on the basis of contracts that paid fixed amounts for each user registered, each uptake regularized, and each meter installed, as well as for each square kilometre included in the register of the secondary network (SF, 1997, p. 19).

In stage two, the contractors were to calculate costs on the basis of the metering of consumers, billing, updating of the registers of users and the network, and acceptance of the reception of duties payable to the city government, among other activities related to commercial aspects of the water supply. In this phase, the payments due to the contractors were to be based at first on a certain amount for each reading and billing, evolving, as more information was acquired, into a formula based on a percentage of the amount charged in each operational zone by the CADF.

In the third stage, the companies were to operate, maintain and rehabilitate the secondary networks of potable water, as well as install systems for detailed measurement and the detection of both visible and invisible leaks. Payment initially was to be by km of pipeline covered, and later by measurement of the water supply in each specific zone. In the general contract, this form of payment was called 'payment by formula'.[2] At this stage, the contractors were to be paid in accordance with a formula that set a price differential between that of the block of water supplied to the contractor and the income generated by retail distribution of the water at the tariffs authorized in the corresponding area. The companies would have to operate and maintain the distribution system. Where they failed to do so, they would lose income through non-payment by users, and their costs would increase where there were leakages: a formula that would promote efficiency and effectiveness. The aim of all of this was to provide an improved service and encourage efficient and rational use, as well as savings of the water itself. At the same time, the price at which water was supplied to the contractor, and the difference between that and the retail price, was to be determined once sufficient information was available on operational costs, the extent of leakages, and levels of payment after a significant time in which the system of unitary payments for specific tasks had been applied (SF, 1997, p. 20).

The other innovation with regard to the private sector's participation in the Federal District was the assigning of the service to four companies rather than one. This was done for social and political reasons that outweighed those of a purely economic nature. The city finance department maintained that, although in theory there could be a different contractor for each of the stages envisioned, efficiency incentives would be fortified by making one contractor responsible for all of them. In that way, for example, the contractor would have a major incentive to make sure the installation of meters was handled correctly in the knowledge that their profitability would depend on their precision. The optimum decision, both for the city government and for the companies, would have been to put one single company in charge of all the stages in one single zone. However, for reasons of a strategic nature,[3] the city was divided into four zones because: (a) should one of the companies be unable to fulfil its contract, any one of the other three would be immediately available to take over, so reducing the monopoly power of each of the contractors; (b) the risks of collusion among contractors would be reduced, given that the greater the number of companies involved, the less chance there would be of them indulging in non-competitive practices; (c) work could go on simultaneously in four of the boroughs where the potential for charging was greatest, thus increasing cash flow; (d) four zones could be assigned without the need to pay a surcharge for the assignation of a zone with less than 250 000 uptakes; and (e) the system would promote the formation of four very solid companies that would increase competitiveness at a national level.

The zones were drawn up using a linear programming model so that the boroughs assigned to each bidder bordered on each other, thus facilitating detailed measurement. The boroughs with the greatest potential were assigned to different groups with the aim of them starting work at the same time, so speeding up the generation of greater cash flows. It was also hoped that the zones would be roughly equal in the number of uptakes and in the projected value of the corresponding contracts. Finally, the priority set by the contracts was the promotion both of sufficient competition among the companies and of the city's development (SF, 1997, pp. 20–21).

As far as the financial aspects of the project are concerned, funding came from the city government's budget. In order to obtain the best possible financial terms, the CADF worked out a mechanism with the National Bank of Public Works and Services (BANOBRAS) that made this bank the agent for the payments of the obligations that the city government undertook as a result of the tender. BANOBRAS promised punctuality in the payments (Casasús, *Construcción*, 1993, p. 31, cited by Martínez Omaña, 2002, p. 182). The financial impact of the change from a system of fixed charges for water to one based on measured consumption would imply an investment of approximately 208 million pesos in 1993 and 1994. In return, this was expected to increase the income of 472 million pesos in 1992, to close to 1 billion pesos in 1994 (CADF, 1993, p. 11, cited by Martínez Omaña, 2002, pp. 182–183). However, as established in the general contract, the contractor had to offer the city government financing for the payments relating to the activities of the first stage. The payments due to the contractor in return for the work that corresponded to that stage were to be made in the form of a bimonthly quota in return for services rendered in regularizing the metering service (Casasús, 1993). At the same time, the contractors were to be responsible for detecting visible and invisible leaks (Martínez Omaña, 2002, p. 183).

Tender and the Choice of Companies

The process leading to private-sector participation was launched in November 1992, when the tender was published. The terms included a stipulation that bids could only be submitted by companies in which a majority of the capital was Mexican. In February 1993, seven bids were submitted and, rather than accepting the most attractive, the decision was taken to divide the city into four zones. The general contracts were signed in the last quarter of 1993, followed by the specific first-stage contracts in May 1994, when fieldwork began. In November and December of that same year, the second-stage contracts were signed. It is important to note that the decision to divide the city into four zones was taken after the companies had submitted their bids and when senior members of the city government had shown concern over the idea of handing over responsibility for the city to one single company because they feared that it would create a monopoly over potable water supply.

In accordance with the new scheme, the 16 boroughs that make up the Federal District were divided into four zones. The division was set by Article 15 of the Federal District's constitution as follows:

- Zone A: Gustavo A. Madero, Azcapotzalco and Cuauhtémoc boroughs, with an estimated 298 557 uptakes.
- Zone B: Benito Juárez, Coyoacán, Iztacalco and Venustiano Carranza, with an estimated 257 825 uptakes.
- Zone C: Iztapalapa, Tláhuac, Xochimilco and Milpa Alta, with an estimated 327 408 uptakes.
- Zone D: the boroughs of Tlalpan, Magdalena Contreras, Álvaro Obregón, Cuajimalpa y Miguel Hidalgo, with an estimated 263 789 uptakes (Table 1).

Foreign private investors were allowed to take part on condition that their investments were considered as such by the law for the promotion of Mexican investment; nor were they given faculties to determine the management and effective control of the consortiums to which they belonged (Martínez de Omaña, 2002, p. 175).

In accordance with the terms of the general contract, each of the companies had to present proposals for the following four areas:

- Client services, including meter reading, billing and attention to the public. This was to be done through offices in each of the boroughs where users could go to apply to be connected or disconnected, change meters, etc. In addition, telephone call centres were to be set up.
- Support, human resources and systems. The responsibility of this area was to provide quality control to ensure that the procedures in use were the best possible at all times.

Table 1. Contractors, including the partners involved, zone, boroughs and number of uptakes, 1994

Zone	Consortium	Partners	Boroughs	No. of uptakes
A	SAPSA	ICA CIE. Generale des Eaux BANAMEX	Gustavo A. Madero, Azcapotzalco and Cuauhtémoc	298 557
B	IASA	Socios Ambientales de Mexico (SAMSA) Severn Trent	Benito Juárez, Coyoacán, Iztacalco and Venustiano Carranza	257 825
C	TECSA	Bufete Industrial BANCOMER Lyonnaise Anglian Water	Iztapalapa, Tláhuac, Xochimilco and Milpa Alta	327 408
D	AGUAMEX	GUTSA Northwest Water	Tlalpan, Magdalena Contreras, Álvaro Obregón, Cuajimalpa and Miguel Hidalgo	263 789

Source: CADF (1994).

- Contracts and technical services. This area was in charge of carrying out the census, regularizing meters and updating the register of networks.
- Operations. This area was to be in charge of meter maintenance (during the second stage) and the pipelines (detection of leaks and repairs) (Martínez Omaña, 2002, pp. 181–182).

Implementation of the general contract was delayed for a variety of reasons (some legal, others administrative, political and financial). Although the winners of the tender were picked in March 1993, the contracts were not signed until September of the same year and were put into effect in May 1994. The main reason for the delay was a judicial complaint lodged by one of the companies that lost out in the tender (GMD/Biwater). Administratively, the GADF needed time to organize itself internally and agree with the contractors on operational aspects such as how the service orders were to be issued and paid. There were financial problems too. The 1994 peso devaluation meant a substantial increase in the cost of the imported meters. There were delays in the installation of meters because the budget assigned by the city government was insufficient to pay for them.

Results According to the Main Players

The perceptions of main players in the Mexico City area were an important part of this research. Hence, interviews were carried out with officials, businessmen, academics and members of the public. In general, those interviewed thought that on the whole the private sector's participation was positive, although as can be seen from what follows, each had a different emphasis. Because they are considered to be the most relevant, only the views of officials (CADF) and of those involved at the level of society in general are presented here (for a detailed analysis, see Marañón, 2004).

Official Views

The broad consensus is that private-sector participation has made a positive contribution to the management of potable water in the Federal District, particularly in the operation of the commercial system (SF, 1997; Zentella, 2000; Haggarty *et al.*, 2001; Beristain-Iturbide, 2002; Saade, 2002; Marañón, 2004). Development of the commercial system was considered part of the first two stages of the general contract signed by the CADF and the companies, including the collection of basic information on the networks, the register of users, installation of meters and the beginning of billing and charging for measured consumption. Improvements in the commercial system can be seen in the register of users, the installation of meters and their maintenance, collection of charges and greater efficiency.

The register of users. The cornerstone of any commercial potable water system is having a users' register that is up-to-date and reliable, something which, as has been seen, simply did not exist in the Federal District. For that reason, at the beginning of the 1990s the CADF put its best effort, with private-sector support, into identifying the users of the potable water service and finding out how many there were. Between 1994 and 1996 a general census was compiled of the properties and water uptakes in all 16 boroughs with the aim of increasing the coverage and updating the ratepayers' roll of the Federal District Treasury. The census became the new register of the city's water users, with updated and specific information on each property and uptake so that each unit of consumption could be billed on a real basis and in a fair manner for each home, apartment or commercial premises and in accordance with the different branches of industry and commerce (see Table 2).

Measurement. Before 1994, measurement of consumption was virtually non-existent in the Federal District. The few meters installed were of widely varying types and brands, and they received no maintenance. In that same year the CADF launched a meter installation programme with the aim of achieving, in the medium term, levels of coverage comparable with those of developed countries. In the first year of the programme, 205 200 meters were installed, rising in the following years to a peak in 2002 of 1 264 500, of which 914 100 are type 'A' and 350 400 type 'B'.[4] As a result, the coverage of detailed measurement has reached more than 90%, a figure comparable with that of developed countries.

Table 2. Evolution of the updating of the Federal District register of potable water users, 1994–1996

	1994	1995	1996	Cumulative
Total number of users (000s)	544.9	1074.2	144.8	1763.9
Register of networks (km² 000s)	83.3	320.4	276.3	735.9

Source: SF (1997).

The CADF maintains that the metering has helped to promote more rational use of potable water supplies by consumers, persuading them to adopt habits that favour savings. Evidence of this lies in the fact that, although the city has received no additional potable water supplies since 1995 (and will not receive them until the completion of a project to supply the city from external sources), there have been no serious interruptions to the service. On the contrary, the meter-installation programme and the programme for the detection and suppression of leaks have enabled the recovery of substantial volumes of potable water. Even so, however, some 200 000 people in the Federal District have no access to potable water and an indeterminate number, especially in the boroughs of Tlalpan and Iztapalapa, receive an intermittent service (Haggarty *et al.*, 2001).[5]

Total coverage of detailed measurement would require the installation of about 400 000 more meters, which is not feasible given the state of the system, the difficulties in gaining access to isolated and remote areas, and the refusal of some users to accept them, among other factors. As a result, the CADF believes that a certain number of uptakes will never be measured. The target to be reached for coverage has, therefore, been set at 95%, which implies installing almost 100 000 meters between 2002 and 2006, in addition to those that will have to be installed as a result of the city's natural growth.

The CADF has paid special attention to major non-domestic consumers, of which there are 16 050 (0.9% of the total register) accounting for 46% of all income. The major consumers are managed by the CADF itself, which reads the meters and calculates consumption levels. The CADF wants to make this process more efficient by introducing a system of remote reading in the medium term, thus raising the major consumers' contribution to 60% of overall income. It has been reckoned that, with that level of income growth, the investment needed in order to modernize the metering system of the major users could be recovered within a year.

The large number of meters now in use means that there has to be a large-scale maintenance programme to keep them in the best possible operational condition, correct any damage or defects that emerge as users point them out, and supervise and control the work done by contractors. For example, between 2001 and July 2003, maintenance was provided for 194 300 meters at an average annual cost of 74 million pesos. The costs were recovered by improvements in the measurement and control of consumption. According to the CADF, the benefits outweighed the costs.

Detection and elimination of leaks. The administration of the centre-left Party of the Democratic Revolution (PRD) that took over the city government in 1998, changed the criteria for the charges to companies, and launched a massive programme for the detection and elimination of leaks, with the aim of solving one of the system's gravest problems— the loss of water from the outdated secondary network. Measures that were taken included replacing asbestos tubing with PVC, which is more durable and flexible and thus less susceptible to fracture on sloping terrain.

According to the CADF, the programme for the detection and elimination of leaks has led to savings of $2.8\,m^3$ per second of potable water, producing benefits for 1.2 million city inhabitants in return for an investment of 1.516 billion pesos. Government critics, however, maintain that the true savings are much less given that the zones to be repaired were chosen without any statistical criterion based on sampling. There were no savings, they say, because although leaks were suppressed on the periphery, the tubes fractured further down the line because of the increase in pressure.

Collection. The CADF mainly collects rights on supply of potable water and discharges to the drainage system. Charges are also imposed for late and overdue payment, fines and expenses in cases of dispute. In addition, the CADF retains the value-added tax (VAT) charged to non-domestic users. In general terms, collection has followed a positive trend, rising by 70.5% in real terms between 1992 and 2001, offsetting the major fall that took place in 1995 and 1996. The key factor behind this development has been the strength of the commercial system, mainly the smooth operation of the meters and the billing and collection. The tariffs themselves have fallen in real terms, above all since 1996 (Haggarty *et al.*, 2001; Saade, 2002; Marañón, 2004) (see Table 3).

Debt recovery is difficult because of the general culture of non-payment of the bi-monthly rights for potable water supply. Some users do not want to pay; others cannot because of their poverty. The culture of non-payment developed as a result of the low, fixed-quota tariffs previously charged by the Federal District Treasury which also had a very limited capacity for debt collection. Traditional methods of debt collection, which proved very costly, are being replaced by mechanisms that improve the cost-benefit relationship between expenses incurred and the amounts recovered. These mechanisms include invitations by letter to pay the debt, instalment plans, and information on the bills that tell users exactly where they stand.

Even so, it should be emphasized that volumetric meters have been located for 72% of domestic consumers, a figure that rises to 90% taking into account the fact that large numbers of apartment buildings—in both upper-class and working-class neighbour-hoods—measure the overall consumption then divide it up among the residents. Payment by fixed quota is marginal and happens only in areas where there is no meter, when meters are broken or when meter reading is impossible. The tariff structure includes several ranges of consumption of which the first is from 0 to 1 m^3 a month, for which there is no charge; minimal charges are made for amounts greater than this. There is also an additional charge for each subsequent cubic metre.

Table 3. Evolution of collection of charges for potable water service in the Federal District, 1992–2002

	Amount recovered (in million pesos)		Real index (1990 = 100)
	Nominal	In real terms (base: 1990)	
1992	471.0	354.2	100.0
1993	572.0	398.3	112.4
1994	712.0	463.1	130.7
1995	769.0	329.1	92.9
1996	1080.0	362.0	102.2
1997	1508.3	436.8	123.3
1998	2053.5	501.4	141.6
1999	2505.0	544.6	153.7
2000	2788.4	556.3	157.1
2001	3159.9	603.9	170.5
2002	3000.0	551.3	155.7

Source: CADF (2002).

Efficiency. The efficiency of the water system has improved notably since 1996 (see Table 4).

Physical efficiency has increased from 62.6 to 68.9%, which means that physical losses of water have dropped noticeably. At the same time, efficiency in the measurement of consumption has risen from 49.1 to 90.2% as a result of the installation of meters and the change in charging from a fixed-quota system. Collection has also improved substantially, increasing from 64.8 to 76.9%, reflecting advances in the register of users as well as in billing and collection. Overall efficiency (measured as a product of the three indicators already mentioned) increased from 19.9 to 47.8%, a notable improvement that contrasts sharply with the low levels recorded at the end of the 1980s (10%) (Beristain-Iturbide, 2002).

To summarize, according to the CADF, potable water management in the Federal District with private-sector participation has led to the achievement of several important results. The authorities now have basic information on the system in terms of the networks and users, metering has been applied in most cases, income has risen, and consumption has been reduced, as has the level of physical losses. On those last two points, Rodarte (2002) maintains that metering has boosted a culture of savings in both high and low-income homes. On the other hand, the same author claims that the high-tech detection and suppression of leaks has led to the recovery of $2.8\,m^3$ of water per second. Rodarte concludes, however, that the initial objective of balancing the system's books has not been achieved because the overall efficiency is only about 50%, which means that only half of what is spent is recovered. Saade (2002) estimates that annual subsidies amount to some $2 billion dollars and suggests that tariffs be increased.

Rodarte (2002) emphasizes that a sustained increase in potable water tariffs is not viable in the Federal District for social reasons, since 72% of the population live in poor areas. When the PRI was the political party in power, ruling the capital under a virtual one-party system until 1998 and the nation until two years later, the city's inhabitants always considered water to be a vital service that has to be provided by the state. The PRI maintained water as a public service and curtailed efforts to privatize it. That concept has been strengthened, if anything, by the arrival of the PRD in power, which has the overwhelming support of the city's poor.

However, more attention needs to be paid to the current management of tariffs. The essential aim of the current tariff structure is to collect money; it is not designed to encourage savings of water. If this situation is to be changed, the ranges of consumption will have to be re-drawn by providing more support to those less able to pay and increasing the rates for the middle and upper classes (Marañón, 2004). In this context, it might be worth considering water as having two aspects, one social and the other economic. As far as the first is concerned, a certain volume of water could be supplied at low cost to meet the basic needs of the poor, while the rest could be sold at prices whose reference would be closer to opportunity costs. This could be one way of balancing the needs of an increasingly poor population with that of modifying users' perceptions on the availability of water.

Society in General

The impact on society of the experience of private-sector participation in the management of potable water in the Federal District must be regarded on two levels. One level is the

Table 4. Efficiency indicators for the management of potable water in the Federal District 1996–2002

	1996	1997	1998	1999	2000	2001	2002
Relative Data							
Physical efficiency[a]	62.6	63.0	64.5	66.3	68.0	69.2	68.9
Efficiency of measurement[b]	49.1	80.0	86.0	88.5	90.3	90.1	90.2
Efficiency in charging[c]	64.8	63.5	76.0	81.7	79.3	83.1	76.9
Overall efficiency[d]	19.9	32.0	42.1	47.9	48.7	51.8	47.8
Absolute Data							
Physical efficiency							
Volume of water supplied to users (millions of m^3)	686.6	690.6	691.9	720.2	752.8	752.2	757.5
Volume of water produced (millions of m^3)	1096.9	1096.1	1072.8	1086.3	1107.0	1087.0	1100.0
Efficiency of measurement							
Number of users billed (thousands)	1477.5	1620.2	1644.0	1681.1	1720.0	1769.1	n.d
Meters installed (thousands)	737.2	1051.6	1137.3	1187.1	1228.6	1255.9	n.d
Number of bills based on meter readings (thousands)	725.6	1260.6	1408.3	1505.1	1552.8	1582.7	1590.0
Numbers of bills issued (thousands)	1478.2	1575.7	1637.6	1701.2	1720.0	1756.0	1800.0
Efficiency in collection							
Amount collected (billions of pesos)	1.1	1.5	2.1	2.5	2.8	3.2	3.0
Amount billed (billions of pesos)	1.7	2.4	2.7	3.1	3.5	3.8	3.9

Source: CADF (2002), non-published information.

Notes: [a]Volume of water delivered/Volume of water produced; [b]Number of bills for metered service/Number of bills issued; [c]Amount charged for water/Amount collected; [d]Physical measurement; Physical collection.

change in quality and quantity of the service, the other refers to the relations between state and society. As far as the first is concerned, the perceptions of the people interviewed (general inhabitants and leaders in the boroughs of Coyoacán, Tlalpan and Xochimilco) were contradictory. On the one hand, there was recognition of the progress made by the authorities in attention to the public through the centres set up for that purpose. The centres make it easier to pay bills, respond to inquiries on issues such as overcharging, make applications for connections and disconnections and register changes in property ownership, and make complaints about poor service or lack of adequate attention by CADF personnel. It should be remembered that the CADF insisted that the companies establish at least one centre for attention to the public for every 300 000 users, as well as having enough space to attend to the complaints, applications, payments and inspections. For the inspections, offices were opened in each neighbourhood so that users' data could be checked *in situ*. At the same time, the CADF insisted that the contractors could acquire and maintain in operation a computer system that would give them access to up-to-date commercial information.

At first, the introduction of charging based on meter readings caused general ill-feeling, particularly among groups of users in poor areas. The problem was that they identified the increase in their bills as an increase in tariffs instead of understanding that it was the product of a change from payment of a fixed quota to payment by measured, or real, consumption. There have been demands for a tariff reduction and improvements to the service, as well as claims of overcharging. Other complaints have centred on the reclassification of the register in neighbourhoods and housing estates, where in some, residents have blocked the installation of meters.

As for the quantity of water supplied, the evaluation is negative, mainly in the areas of the city where supply is intermittent. Indeed, it appears that there are many more such areas than has been officially recognized.[6] The timing and duration of the intermittent supply is variable. The supply is switched on only once every three, four or six days, or even once a fortnight, and then for only two or three hours at a time. Intermittent supplies are applied in poor outlying areas, squatter settlements or in high-altitude zones where there are pressure problems. People who live in these areas complain about the inequality in distribution of potable water and oppose the metering system with the argument that they are charged for the air that passes through the meters and not for the water. They argue that they should be charged a fixed quota and not by volumetric consumption. Haggarty *et al.* (2001) maintain that problems of service quality persist, including poor water quality and intermittent service, particularly in the south and east of the city. It is worth recalling that the reliability of the water service remains the responsibility of the boroughs. In 1998, residents of nine of the 16 boroughs suffered routine cuts to the service, although the severity of the problem and the number of people affected varied considerably. In these areas of the city, water quality is extremely poor, due in part to the fact that the underground water contains high concentrations of magnesium and also because the south-east of the city is the last to receive the water that comes from the Cutzamala System.

Another aspect worth noting was that the users who were interviewed, irrespective of their social class or whether or not they were leaders of social movements, did not know about the private sector presence in the management of the service. Those interviewed were aware of improvements to the service but had no idea of the companies' positive contribution. Hiernaux-Nicolas (2002) maintains that, considering the high degree of politicization of the inhabitants of Mexico City, it must be concluded that the issue of participation of risk capital has been kept out of public debate and the population has not been informed of the companies' role, perhaps through fears that such knowledge could lead to protests, as it did recently in Cochabamba, Bolivia.

According to Martínez Omaña (2002), the institutional arrangement in force in the Federal District since the mid-1990s consists in the participation of the private sector through service contracts, with the state maintaining control over property rights in the infrastructure, responsibility for provision of the service, and the power to fix tariffs. The state does delegate certain basic responsibilities to the companies, including the design and distribution of bills, measurement, collection, updating of the register of users and management of the centres for attention to the public. However, the companies perform these services in the name of the CADF, not in their own name as would be the case of a concession or an asset sale. This is a reason why users, no matter their social class, are unaware of private-sector participation. Within the current arrangement, the CADF ought to be, as was established on its creation, the sole authority in charge of water management in the Federal District, co-ordinating the functions of the DGCOH and the boroughs with private-sector support. However, this idea failed to materialize. Until 2002, the CADF managed the contracts with the companies and their regulation, as well as supervising all the activities assigned to the private sector, principally in the commercial area. The DGCOH was in charge of carrying out the programmes to expand the infrastructure and the operation of the primary networks, while the boroughs were responsible for maintenance of the secondary networks of potable water and drainage, the repair of uptakes and the repair and maintenance of pipelines, as well as administration of intermittent supplies in areas of water shortage.

As far as the relation between state and society is concerned, it is important to note that the new strategy made no attempt to change the vertical, bureaucratic nature[7] of present-day water management in which decisions are centralized, with no participation by the public, and an abundance of red tape. The result is a substantial gap, between users and the authorities, in the perception of the social and geographical realities where water is in short supply. Efforts have been made to bridge the gap through the use of social representatives whose job is to agglutinate residents' demands, although some of them have personal agendas that they pursue.[8] The distance between the two sides and the absence of communication simply adds to users' ignorance of the current institution's accord. The popular view is that the authorities are fragmented: one (the CADF) that takes care of the commercial aspects (metering, billing and collection), and another (the boroughs) that is involved in distribution of the water, management of the interruptions to the service, and repairs to leaks, and hence always being criticized for shortages. Moreover, this view of a division corresponds to reality. It was not until 2002 that a new body was created, the Mexico City Water System (Sacmex), to take charge of the system by merging the CADF and the DGCOH, although the boroughs clung on to their responsibilities. Until Sacmex was founded, the hallmarks of the Federal District were a scattered approach at the institutional level and the lack of a common, integral vision of the problems and possibilities for water management.

Concluding Remarks

Throughout this paper, a detailed revision has been made of the process of private sector participation in the Federal District's water management between 1992 and 2002.

This aspect is of key importance because the contract expired at the end of 2003 and the parties were negotiating the terms of a new arrangement. As a result, it is important to bear in mind an evaluation based on experience.

The consensus view of the new strategy is favourable among the main players, including businessmen, academics and representatives of the institutions involved and society in general. The main advantages have been the establishment of updated registers of the networks and users, the mass installation of meters, the change from fixed quotas to measured consumption and the increase in the volume of consumption and of collection in real terms. The same positive view is held of attention to users, thanks to the offices for that purpose and the call centres set up by the companies.

Regulation by the CADF was adequate, but solutions have yet to be found for the problems experienced in co-ordinating the efforts of the authorities (DGCOH and the boroughs) and the private companies in order to define the attributions of each of them in the repairing of leaks. On occasions when users have made specific reports of a leak, all three parties have turned up to solve the problem, duplicating functions and efforts.

Uncertainty surrounded the future of private sector participation in the potable water service of the Federal District as the contract drew to a close in August 2003. The companies lacked information from the city government on the immediate future (whether private sector participation was to be maintained at the same level or increased, or whether responsibility might be put back in state hands). At the end of 2003, both sides were negotiating a new agreement that seemed to respond to the general concerns. The government was seeking to maintain the private sector's participation in the tasks already assigned to it, but was offering lower rates for each task and the same rate to each of the companies. At the same time, the government was also proposing to extend the ambit of risk capital by applying it to the management of the 'major consumers' and giving it powers to cut off the service to non-domestic customers whose payments fell overdue, as well as taking action to recharge the aquifers. Responsibility for the water supply was to remain in state hands with the private sector continuing to participate in specific areas without altering access to water as a citizens' right.

The importance of highlighting the Federal District's proposal for the articulation of the roles of the state and the private sector lies in the fact that the private sector takes part through contracts, providing technology, organization and international experience. For its part, the government decides on volumes and delivery times, as well as in the operations carried out in the name of the water authority. The social nature of the service is retained, since the government keeps control of the tariffs. This is relevant in the light of current international trends in the relationship between public and private sectors in water management. These are marked by an ideological weakening of the pro-privatization tendency and a decline in investment by the major transnational companies due to the poor results recorded at a social level in several Latin American countries and growing protests as a consequence.*

Notes

1. When the invitation was sent out to the private sector, there was practically no information in terms of the length of the network, the roll of users, structure and volume of consumption, levels of physical loss, billing and collection and costs, among other things. As a result of the lack of necessary financial and operational information, negotiation of the contract on a reasonable basis was simply impossible, according to one of the businessmen interviewed (Mexico City, August 2003).

2. The manner in which contractors were to be paid in each stage is described in The General Bases of Tender, clause four, section 4.6. Formula payments are described on pp. 34–35, subheading 4.6.3.1.

3. Number 5 of the subsection of Chapter III of the General Bases of Tender.

4. The collapse of the peso led to a 119% increase in the cost of type 'A' meters and 204% for type 'B'. As a result, the 1995 target for installation of meters was cut back from 586 700 to 219 300. The plan contemplates the installation of both velocity and volumetric meters. The mechanism of the velocity meters (type 'B') makes them more resistant to the suspended solids in water than those of type 'A'. The type 'B' meters are cheaper and the original aim was to install them in areas where the recovery of the investment in the more expensive apparatus appeared to be problematic (SF, 1997, p. 60).

5. According to a study by the Autonomous Metropolitan University (2000), some 12% of homes in the Federal District receive intermittent supplies of water and about 2% have no access to potable water.

6. According to borough officials in charge of potable water provision, a greater proportion of the population receives intermittent service than that declared by the DGCOH. The DGCOH, in assigning a volume for a particular zone, assumes that all the residents there receive water. Once the water reaches a neighbourhood, however, pressure is sufficient only to bring it to a fraction of all homes.

7. On this point, see Coulomb (1993).

8. On this issue, see Treviño (1999) and Avila (2003).

References

Avila, P. (2002) Water, power and conflicts in a middle-size city, in: P. Avila (Ed.) *Water, Culture and Society in Mexico* (Michoacán: Colegio de Michoacán/IMTA).

Beristain-Iturbide, J. (2002) Policies for commercialization and public expenditures for water. National Forum on Legislation for the rational use of water, Federal District Congress, México, DF, 28 February.

CADF (1993) *A New Strategy for Water for Mexico City* (México, DF: CADF).

CADF (1994) The new strategy on water for the Federal District. Presentation to the Head of the Department of the Federal District (México, D.F.: CADF).

CADF (1995) *Water Commission of the Federal District and the New Strategy* (México, DF: CADF).

Casasús, C. (2002) CADF, general bases of tender, general contract, 1993, in: M. C. Martínez-Omaña (Ed.) *Private Management of Public Service: the Case of Water in the Federal District 1988–1995* (Mexico, DF: Instituto Mora, Plaza y Valdés Editores).

Coulomb, R. (1993) The participation of the population in the management of the services for urban areas: ¿privatization or socialization?, in: A. Azuela & E. Duhau (Eds) *Urban Management and Institutional Changes* (Mexico: Mexican Autonomous University).

Haggarty, L., Brook, P. & Zuloaga, A. M. (2001) *Thirst for Reform? Private Sector Participation in Mexico City's Water Sector* (Washington, DC: World Bank).

Hiernaux-Nicolas, D. (2002) The water agenda in Mexico City: challenges and perspectives, *Sustainable Urban Services*, Seminar, Santiago de Chile, 10–11 July 2002.

Marañón, B. (2004) Tariffs for drinking water in the Metropolitan Area of Mexico City, 1992–2002, in: C. Tortajada & A. K. Biswas (Eds) *Water Pricing and Public and Private Participation in the Water Sector* (Mexico: Editorial Porrúa).

Martínez-Omaña, M. C. (2002) *Private Management of Public Service: the Case of Water in the Federal District, 1988–1995* (Mexico DF: Instituto Mora, Plaza y Valdés Editores).

Ministry of Finances (SF) (1997) *Structural Change in the Water Sector in the Federal District 1992–1997* December (Mexico City: SF).

Rodarte, L. (2002) Regulatory framework for water service operators, *Sustainable Urban Services*, Seminar, Santiago de Chile, 10–11 July 2002.

Saade, L. (2002) Challenges in the water sector in the Mexico City Metropolitan Area. Sustainable Urban Services, Santiago de Chile Seminar 10–11 July.

Treviño, A. (1999) Actors and organizations for water, *Ciudades*, 43, July–September.

Zentella, J.C. (2000) The participation of the private sector in the management of the private sector of water management in the Federal District. Economic, technical and administrative evaluations, 1984–1996 MSc Thesis, CEDDU, Colmex.

Participation of the Private Sector in Water and Sanitation Services: Assessment of Guanajuato, Mexico

RICARDO SANDOVAL-MINERO
State Commission of Water, Guanajuato, Mexico

Introduction

The Administration of Ernesto Zedillo (1994–2000) established a Financing Fund for Infrastructure (FINFRA), with the support of the National Bank of Public Works and Services (BANOBRAS) by means of a trust fund whose main objective was to promote private investment in specific public service projects. Several treatment 'turn-key' projects were funded by FINFRA using private sources of financing, however the administrative procedures were too cumbersome and the projects experienced many delays. At the beginning of the current Administration, a new version of the fund (FIFRA 2) was introduced, together with a new Program for the Modernization of the Water Operating Agencies, also known as PROMAGUA. This programme applies differentiated subsidies and promotes private participation in a wide range of schemes with the specific objective of enhancing the administrative skills of the operating agencies.

Within this context, the paper will explore the feasibility and the actual need of resorting to private investment in Guanajuato, based on the current drinking water, sewerage and sanitation state indicators. To that end, the assumptions used to create the various existing private investment programmes will be reviewed in order to design a proactive strategy that will eventually lead to an efficient implementation of this type of financial schemes.

Framework in Guanajuato

Guanajuato is located in the central region of Mexico. It occupies 1.56% of the country's surface area, and it has almost 5% of Mexico's population and only 0.7% of the available water supply. Due to its intensive dynamic development and the institutional lack of resources, Guanajuato's natural water availability has reached 35% of over-exploitation, resulting in 2 m of annual drawdown of its aquifers. Apart from this unbalanced situation, the public and industrial supply systems share the same deficiencies and problems that are common to the rest of the country: lack of quality, global efficiency and coverage. It should also be noted that in Mexico the water sector has developed historically under a highly centralized regime, where most of the regulatory, operating and infrastructure developments were in the hands of the federal government, with the exception of the drinking water and sewerage systems that were passed on to the municipalities in 1982. That is why there are still many gaps at the state and municipal levels with regard to their institutional, financial, human capital and social value capabilities. The most recent actions stemming from the state water programme are focused on solving the following major challenges: to balance the supply-demand-quality equation and to achieve efficiency, quality and coverage in service as well as in institutional support.

Guanajuato has 8936 localities (INEGI, 2000) distributed in 46 municipalities, 12 of them have a population of over 50 000, with a population of 2.2 million in total. This represents 48% of the total population in the state. Another 99 localities have populations of between 2500 and 50 000 and they represent 19% of Guanajuato's population. This means that there are 8821 rural localities inhabited by 1.5 million people suffering from very low levels of water coverage (Table 1).

Looking at this information, the current Urban Sustainable Management Strengthening Programs must be directed at the creation of rural infrastructure development subsidies. Water and sewerage coverage is relatively good in urban localities (better than the domestic average) while rural areas still face major inadequacies, mainly in sanitation.

These assumptions were the foundations of an ambitious programme created in 1995 that aimed to furnish the municipal agencies with enough tools to achieve sustainable growth and strengthen their position as public entities. The present state Administration has given continuity to that strategy.

By 2001, an overall estimated efficiency of 49% had been reached in 46 municipalities affecting 2.8 million people, with a micro-measurement coverage of 52% and a drinking water purification level above 96%. At present there are 35 operating agencies acting as

Table 1. Stratified service coverage, Guanajuato

Stratum	Number of localities	With water	With sewerage	Household occupants
>50 000 people	12	97.37%	95.82%	2 213 569
2500–50,000 people	99	96.73%	85.89%	891.479
<2500 people	8821	81.55%	39.30%	1 520 882
Total*	8932	92.05%	75.32%	4 625 930

Note: *This information refers to people living in private homes; it does not include shelters.
Source: Calculated by CEAG (Comisión Estatal del Agua de Guanajuato) with data from INEGI, Housing and Population Census 2000.

decentralizing units of the municipal government, with Citizen Boards. The CEAG and the operating agencies have carried out 37 Rate Studies, 42 User Registers, 42 Commercial Systems, 41 Cadastral Surveys of drinking water and sewerage systems, as well as 13 Master Plans that had to be deferred until all the other elements were complete. Table 2 shows a detailed abstract of the state figures per population stratum.

It is important to note that the state averages do not reflect the discrepancy found in the cities of Guanajuato. According to a current survey (CEAG-LARUZ, 2002) conducted with a sample of nine cities, the accrued physical and accounting losses found were the following: Abasolo: 39%; Acambaro: 34%; Celaya: 19%; Ciudad Manuel Doblado: 51%; Comonfort: 53%; Capital City of Guanajuato: 9%; Moroleon: 37%; Romita 49% and Salvatierra: 38%. Figure 1 represents the ratio between the billed volume and the volume of production, as reported in 2001 by 26 of the operating agencies.

The information presented in Figures 1 and 2 is heterogeneous because sometimes the calculation of billed volumes is affected by the addition of the different consumption ranges, as well as by the lack of congruency between production and billing dates (backlogs which are billed afterwards). Micro-measurement coverage increased by 62% in 2002 (CEAG-LARUZ, 2002). These estimations are based on the number of installed devices, since human measurement errors, operating and installation flaws have been detected in almost 40% of the metering devices. Figure 2 includes the micro-measurement data reported by 24 operating state agencies.

These data and the Figures show the great diversity of operating conditions that Guanajuato agencies have to cope with. It is also true that each agency has reacted differently depending on the access that they have had to the state resources for consolidation, as well as the quality of their relationship with the municipal officials: the extent of their financial and administrative independence, and the level of interest and participation of their communities, which are eventually translated into different degrees of collection and users response.

In addition, the diversity of circumstances also creates a diversity of conditions, in a struggle to reach a sound sustainable operation, despite the financial problems of each system that will be described later. It must be stressed that the cities of Guanajuato, León with almost 1 million people, Celaya and Irapuato with almost 400 000 people, Salamanca with almost 200 000 and the 15 remaining medium-sized cities with more than 20 000 people, are good examples of the conditions prevailing in other regions of the country. With the exception of Guanajuato, the capital city, where almost 40% of the water supply comes from two dams, and León, which takes 3% of its water supply from a small embankment, the other cities totally depend on groundwater supply (suffering from an intensive drawdown that results in the more frequent presence of natural contaminants, such as arsenic and manganese). Another important aspect to take into account is the outstanding political plurality that Guanajuato has enjoyed for over a decade.

Financial Problems

According to the National Water Commission (CNA), an annual investment of 17 billion pesos is required to reach and maintain 96, 89 and 65% coverage of drinking water, sewerage and wastewater treatment services during the next 25 years. Currently, 3 billion pesos are invested by the federal government and the agencies. An additional annual investment of 5 billion pesos is needed to maintain and operate the system. However, on

Table 2. Consolidating actions and situation per population range in Guanajuato

Range	No. municipalities	Total population	Technical-operating consolidation						Administrative consolidation	Financial/commercial consolidation		
			Potable cadastre	Water sewerage cadastre	Master plans	Efficiency	Micro-measurement	Disinfection	Decentrilized agencies	Rate studies	Users register	Commercial system
100–2499	5	6601	5	5	0	nd	0%	79%	0	1	1	1
Subtotal/rural	5	6601	5	5	0	nd	0%	79%	0	1	1	1
2500–4999	5	19 327	5	5	0	nd	31%	95%	1	3	5	5
5000–9999	4	25 569	4	4	1	20%	51%	99%	3	3	4	4
10 000–14 999	3	37 843	3	3	0	61%	63%	92%	3	3	3	3
15 000–19 999	3	51 115	3	3	1	59%	90%	98%	2	2	3	3
20 000–49 999	14	417 495	14	14	6	47%	53%	95%	14	12	14	14
Medium-size cities	29	551 349	29	29	8	47%	54%	96%	23	24	29	29
50 000–99 999	8	480,039	6	6	4	51%	77%	96%	8	8	8	8
100 000–499 999	3	733 898	1	1	1	46%	30%	94%	3	3	3	3
500 000–999 999	0	0	0	0	0	0	0	0	0	0	0	0
1 000 000 – y más	1	1 020 818	0	0	0	57%	100%	100%	1	1	1	1
Large cities	12	2 234 755	7	7	5	50%	67%	96%	12	12	12	12
Subtotal/urban	41	2 786 104	36	36	13	49%	58%	96%	35	36	41	41
Total	46	2,792,705	41	41	13	53%	52%	96%	35	37	42	42

Source: State Water Information System, CEAG (2002).
Notes: Calculated from information provided by the water utilities and the General Census on Population and Housing, by the National Institute for Statistics, Geography and Informatics (INEGI, 2000).

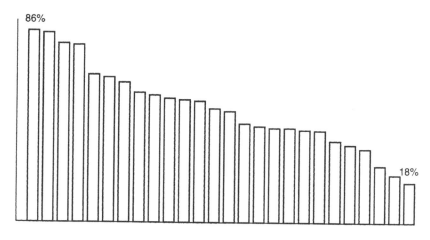

Figure 1. Billed volume divided by volume of production. *Source:* State Water Information System (CEAG). *Note:* Data provided by 26 operating agencies in Guanajuato, 2001.

average Mexico's users pay for only 300 litres of each 1000 litres that are being produced in the country.

Guanajuato's situation is quite similar. Taking into account all the investment required to increase coverage, plus the investment required to achieve the systems consolidation, CEAG has calculated that from 2000 to 2006, 2 billion pesos will be needed. During the first two years, the state government has invested 300 million. The operating agencies and the state municipalities invoice for approximately 900 million pesos a year (CEAG-LARUZ, 2001). Almost half of this amount is billed by the operating agency of León and it does not include any commercial or physical losses, which would leave their actual collection at almost 630 million pesos (Figure 3). Therefore, the annual investment capacity in Guanajuato is estimated at less than $240 million pesos and it is concentrated

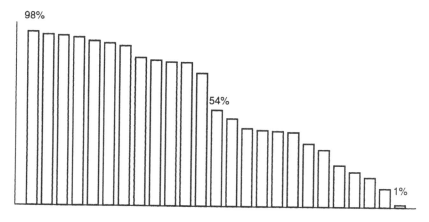

Figure 2. Installed micro-measurement percentage. *Source:* State Water Information System (CEAG). *Note:* Data provided by 24 operating agencies in Guanajuato, 2001.

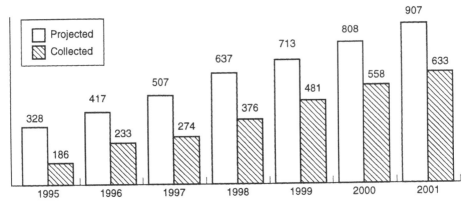

Figure 3. Forecasted collection versus actual collection (in million pesos, 2001). *Source*: CEAG-LARUZ (2002).

in the 11 agencies that are capable of investing to improve and expand their infrastructure (CEAG-LARUZ, 2002).

In addition, recent estimations from CEAG have calculated a total investment requirement of $160 million dollars to rehabilitate the infrastructure that is either too old or has been damaged by underground failures, since there have been cases of tectonic subsidence and over-exploitation in some of the main cities of Guanajuato lowlands (Bajío). The existing infrastructure has also been partially damaged by project or operational mistakes; some wells need refitting and in some cases need to be relocated due to the intensive exploitation of the aquifers in the area. In addition to these figures, investments are required for the large supply projects for some cities in Guanajuato (5 million for San Nicolas Dam in Rio Verde; and 250 million for Rio Santa Maria Project), and almost $50 million for the construction of 20 water treatment plants that are urgently needed in the state. This is apart from all the money needed to strengthen the organizational side of the system. It becomes quite obvious that more financial resources are needed in the state to improve its hydraulic system. New capital sources of finance must be opened along with the development of sustainable cost recovery capacity. This paper will deal later with cost recovery as a component of social financing within Guanajuato's technical and legal rate structure.

Operating and Financial Risk Analysis for a Sustainable Operation of the Existing System

Finance for Water Development vis-à-vis Private Investment

Among the different sources of finance for hydraulic development, private participation as a 'responsible supplier' delegate entity should be considered as an alternative within a wide set of characteristics.

The Table 3 (Mould, 1980) shows a classification of finance mechanisms that can be used to suit different environments and have been taken from different sources. Although concrete mechanisms of private investment are not mentioned in the Table, it can be used as a framework.

From Table 3, it can be observed that Mexico's water and urban sanitation hydraulic development is mostly based upon local sources in the form of domestic allocations, loans and very little bilateral assistance, with the incipient issuance of municipal bonds and other equity instruments. According to Table 3, some service and concession agreements have become external loans requesting the submission of collateral by the municipal, state or federal government, at least in the form of subordinated capital (to recover the infrastructure's useful life). Quite clearly, the financing alternatives are still a rich source of investment diversification. However, the 'left side' of the equation should not be neglected—the assets that form part of any organization, in this particular case the water and sanitation system that needs access to finance, no only in terms of its physical or monetary capital, but also in relation to its stability and income flow. This simple logic considers debt capacity as the first condition to access any source of finance and therefore it 'twists' when private investment is considered as an objective more than a means. In this manner the agreements are subject to a series of commitments regarding rate fixation, regulations and collateral submission, which in the strict sense should have come first and set the conditions for private investment feasibility, instead of the other way around. There will be another look at this point

Table 3. Sources of finance for hydraulic development

Sources of finance	Loans, bonds	Contributions, donations	Types of finances Shares, participation	Internal cash production	Users contribution
Local operating agency					
Local government (municipal)	B	B		A,B	A,B
Local consumers or investors	B		B		A,B
Domestic					
Federal or state government	A,B	A,B	A,B		
Capital and equity markets	B				
Commercial and development banking	A,B		B		
International					
Capital and equity markets	B		B		
Bilateral assistance	A	A			
Export loans	A,B				
Foreign governments					
Multinational assistance					
Combination of foreign governments	A	A			
Multilateral agencies	A				
Joint-financing agreements among bilateral and multilateral agencies	A				

Note: A = applicable to developing countries; B = Applicable to developed countries.
Source: Mould, 1980.

later in the paper.

Financing Management of the Operating Agencies

In those countries where water operating agencies or their peer institutions have access to capital markets and to commercial loans, their skill in identifying and properly managing future risks that may affect the cash flow of the system, depending on their financial and operating leverage, becomes a crucial point of balance which eventually becomes their main source of creditworthiness. In the Mexican context, there is no sense in talking of 'financial management' at the water utilities in the terms already explained since most of these agencies depend on non-refundable contributions, on permanent subsidies for capital asset investment, and on local subsidies to keep their rates below the actual cost of water, and only to a minor extent on their internal resources, which are affected by variability of rates and personal judgements within their three main bodies: the Board of Directors, the Municipal Officials and the State Congress. Very few of these agencies are capable of obtaining a loan from the development banking system and even fewer can think about the possibility of issuing financial instruments (with the exception of León system in Guanajuato).

However, for comparison and programme reference purposes, a series of financial risk management criteria are listed that any agency working within a well-developed financial system should follow (Adapted from Amatetti, 1994). The following strategies aim to minimize the uncertainty of capital investments and expense projections and to optimize capital costs and improve the income stability of the agencies, or a combination of all the above.

- It is important to anticipate and quantify the impact of future regulatory costs. Mexico has undertaken radical changes in water quality, sanitation and fiscal regulations. As an example, during 2002 the electric bills were increased by 6%, in addition to the rates index that was enforced in January, the substitution of the 'Zero rate' by the 'Value Added Tax Water Exemption' that impairs water agencies to credit and recover this tax; and the implementation of a condoning programme of overdue bills and the opening of 'a new account' for the collection of duties. All of these changes were public political decisions that despite their potential benefit or implicit logic, were implemented *after* the time in which the operating agencies prepared their budgets, which resulted in an income decrease, and/or an increase of their expenses of almost 25%, that was reflected in their cash flow on accrued basis. In addition to all these problems is the constitutional exemption made on federal public assets, which are amongst the main users of the system in almost every capital city of the country.
- To avoid limited access or delays in capital investment finance. A delayed operation may result in an income reduction. The lack of public funding in Mexico, together with increasing competition for fiscal resources is introducing high risks implied by the delay that priority urban supply projects are facing, sometimes exacerbated by the unnecessary complexity of procedures.
- To manage the financial risk introduced into the system by the increasing *loan debentures*, in the form of financial leverage and collateral. The crisis in

December 1994 affected the cost of money and eliminated the viability of the existing private investment schemes, drastically increasing the debt that the system had with the development banking. This is a clear example of risk being introduced into a financial system full of self-variability that is typical of an emerging economy exposed to frequent changes in the equity and exchange markets.

- To avoid as much as possible any risk derived from *unsuitable planning* that can turn complex technological costly investments into obsolete projects. A lower demand than that forecast will deplete the system's income and can be fatal in those cases where expenses are taken as fixed expenses in Agreements that consider them as the minimum guarantee that investors have the right to (e.g. the water treatment plant of Puerto Vallarta, in Jalisco state).

- To optimize *maintenance and replacement* of infrastructure and equipment, avoiding an excess of investment caused by an unnecessary anticipation, or having to face extraordinary unforeseen costs due to negligence in maintaining or replacing the assets. In Mexico, the inefficient infrastructure and the failure or poor state of the operating equipment demands a vast amount of investment. Some of the reasons lie in the low performance of the rate system, the omission of the depreciation concept and the political bias towards investing in infrastructural expansion works more than in maintenance. This is an ongoing source of risk that already affects the availability of budgetary resources and the internal cash revenues of many agencies (the syndrome of putting out fires).

- Minimizing the *uncertainty of future revenue*, which may come from the difficulty of obtaining new economically feasible source alternatives, or from sudden changes in the macro-economic variables; changes in the costs of inputs that cannot be reflected immediately or on a temporary basis in the rates; drought restrictions in areas suffering from chronic water shortages; and other factors which reduce the volume that goes into the market or that affects the users willingness-to-pay. In addition, in Mexico it is common to have the municipal officials or the Congress representatives trying to modify the rate structure, or to create pay exemptions and even deviating resources to take care of cash problems in local treasuries.

- To properly manage the operating and financial leverage of the system, deferring as much as possible capital investments with external funding (without falling into the trap of lack of supply) and minimizing fixed operating costs (without leaving unattended the operating processes or overloading the staff and equipment). An agency that has to cope with revenue uncertainty and revenue variability at the same time as trying to pay for higher loan liabilities bound to a contract or to collateral, will incur into a higher financial risk.

- It has been mentioned that those agencies incapable of proving that they have control over the financial risk factors do not maximize their qualifications and gain access to loans and to other financing instruments. According to this brief review, the Mexican operating agencies have limited control of those factors that affect their revenue stability. Their plans do not include the optimization of their investment and they are exposed to frequent cost increasing macro-economic and regulatory changes. The inclusion of rates as duties that must be officially published as part of the municipal income legislation, and must be previously authorized by the State Congress every year, mean it is very difficult to make modifications to the structure and also the amounts of the service rates during intermediate terms. This situation increases the system's vulnerability and the constant cost changes affect the financial position of the operating agencies. In light of this context, it should be questioned whether a private participation scheme per se could give sufficient support and 'isolate' an agency from its sources of financial risk. If the answer is negative, then it should be questioned whether such a risk is then transferred to the state and federal treasuries that are the collateral providing entities.

Private Participation as a Source of Financing

Normally, programmes that promote private investment in water and sanitation services are designed to identify the potential risks that they pose to private investors, creating the regulatory mechanisms and guarantees that will support the risk management or the risk minimization processes, ensuring at the same time some benefits and guarantees of service for the users. From the beginnings of the 1990s until the more recent PROMAGUA programme, the logic shared by the different financing schemes has been based on the assumption that private participation can simultaneously solve two core problems of the developing countries' systems: on the one hand their performance deficiencies, and on the other hand the lack of financial funds. That is why private investment has been regarded as an objective and never as a means.

By exploring the differences in perception it is possible to understand this situation more clearly. The risks that private investors or their promoters have to face in supplying water and sanitation services can be divided in four types (Idelovitch & Ringskog, 1995):

- Commercial, due to inadequate rates or low demand.
- Financial, pertaining their economic stability and its effects on convertibility.
- Technical, due to poor identification of the existing infrastructure.
- Political, derived from obstacles that impair the establishment of suitable rate fixation processes, as well as the lack of legal or social certainty that may impact the assets ownership.

Allegedly, private investment provides higher efficiency because it focuses on yield maximization and therefore can supply more users with better quality and lower costs. According to this logic, private investment becomes the means to achieve the goals, yet, if it is assumed that such goals are part of the nature of private investment, then the implementation of institutional reforms based on a wide political commitment and on public and private consensus about the establishment of clear operating rules and objectives is duly justified.

Promoters of private investment must manage the risk involved for each *party*, namely the investors and the users, placing between them a regulating entity completely independent from both parties and from the government itself. Brook-Cowen (1997) presents a list of requirements for a successful implementation of private investment options. The requirements differ in importance according to their corresponding option, and they ensure a potential benefit with each scheme. As an example, the following list describes the ideal scenario for the implementation of private investment. Clear support is given by the stakeholders (local government, users) and the scheme's implementation has political commitment.

- The rates allow for cost recovery.
- There is enough information about the system.
- There is a well-developed and applicable regulatory framework.
- The 'receiver' has been classified as low risk country.

Before continuing, two questions can be raised:

- If public agencies can guarantee good service in this industry, together with political commitment and respect, competitive rates, sufficient planning and information, adequate regulations and economic stability, without the users' participation ...
- Then, is there a need for private investment in an industry that has reached all these ideal conditions or should private investment be considered as another option?

Since the former ideas are the essence of this paper, further details will be presented at the end.

Preliminary Evaluation of Private Investment Feasibility in Guanajuato

In order to establish a constructive type of strategy, although it might seem somehow preliminary, the general set of conditions that prevail in Guanajuato should be taken into account in order to determine the requirements for promoting private investment in this industry.

As has been previously mentioned, the commercial environment presents a great disparity among structures, tariffs, amounts, regulatory basis and the enforcement of rates. According to the amendments made to Article 115 of the Federal Constitution which were translated to Article 63 of the State Constitution, at present the rate structure forms part of the Municipal Income Act issued by the State Congress. The new Income Act of Guanajuato required the quick approval of 46 income regulations in a very short time, with the consequent mistakes and omissions that are affecting the agencies' revenue today. A positive factor to be taken into account is the definition of Articles 55 to 65 of the State Water Act, with regard to the concepts and conditions used for rate determination.

The particular legal framework of this service in Guanajuato forms part of Article 117, Section III of the Local Constitution, exhibit a, which confirms the municipal capacity of supplying drinking water and sewerage services. The Municipal Organic Law ratifies this capacity and includes, among other related subjects a detailed description of the conditions and characteristics for the concession of Municipal Public Services. Sections two and seven in Article 161 of the Organic Law are a matter of polemics, since Section two allows for a review of the Concession-Certificate and Section seven allows for a rescue by indemnification. The State Congress is authorized to overrule any agreement, regardless its duration and the term of the current municipal administration.

With regard to the potential technical risk, the cadastral surveys have lessened the probability of having to face unforeseen investment needs. These circumstances, together with the implementation of master plans and specific projects that aim to develop and refit the networks, favour the implementation of short-term agreements with private companies to carry out specific improvement actions and processes. The economic stability and the judicial safety of Guanajuato are comparable or even superior to the average in Mexico. Guanajuato is among the states with lower criminality rates and higher occupational levels. Therefore, it can be concluded that even with schemes that are somehow biased towards minimizing the risk for private investors, Guanajuato has positive conditions to promote private participation. This is another strong reason to look for a strategy that stretches far beyond the concept of considering private investment as a goal.

Service Improvement and Private Investment: Concurrent Strategies

Every city in Mexico is facing performance and financial problems. However, these problems are the reflection of a legal, political and social framework that will not be easily modified if the only approach is the administrative scheme. Core problems must be resolved by a comprehensive development strategy, with or without private participation. Guanajuato's comprehensive hydraulic development programme was created with that objective in mind. It departed from the premise that since the state government in Mexico is not empowered to make significant water-related decisions, then it can act as a supportive link by promoting the systems' permanent improvement and establishing efficient regulatory mechanisms that will foster the continuity of good performance practices in the form of financial, technical, legal and political support given by the state to the municipalities.

Another assumption recognizes that many of the basic administrative system deficiencies can and *must* be identified as well as solved by the state authority, using its fiscal resources to widen the technical and financial scope of the systems, regardless the considerations made about private investment, and even without them. As a continuation of policies from 1995 that were looking to provide new tools to strengthen the business sector (networks, cadastral surveys, rate studies, user registers, commercial systems, master plans, etc.), a continuous improvement programme is currently being implemented to ensure the dynamic and practical application of those tools. The programme is based on a general diagnostic that quickly identifies those areas requiring immediate improvement with high profitability potential. In this manner, the areas that are currently receiving the state's support are also yielding revenue to the sector and are removing the burden of unnecessary expenses to the operating agencies (including troubleshooting and electromechanical rehabilitation of the supply sources, sectorial breakdown, macro and micro-measurement, water recovery, rate restructuring, regulations etc.). In addition, CEAG provides technological support (through a Research Fund that has been implemented with the aid of the State Board of Science and Technology), training and occupational certification, water culture development and legal aid.

These actions do not pretend to weaken the viability of private investment. On the contrary, the intention is to prepare the ground for private investment to be considered as an alternative, not as a way out by the agencies. It cannot be denied that private investment in addition to a sound system can result in key additional benefits (Henderson, 1993).

- It is necessary to change the short-term priorities of politicians for the long-term needs presented by the professional business approach.
- The public sector must have access to financial resources that are currently out of reach.

- The quality of water and service has to improve according to the financial and technical capacity available.
- Staff working conditions must be improved and training provided, as well as recruiting the best people for public service and even act as potential partners of private companies.
- A healthy financial market must be fostered, since it will provide creditworthiness incentives to private companies that wish to remain profitable.
- The approach must be customer-oriented, instead of the current technical service approach that is only taking care of the government needs.

In this manner, a long-term successful profitable strategy will 'release' itself from the need of neutralizing the sources of income uncertainty and from 'locks' that it has been forced to use in order to avoid political intrusion in its administrative affairs. Such a strategy will also be free from expensive and risky collateral, as well as from the already taken for granted mechanisms used in settling disputes, turning into a pure strategy of constructive work based on clear interaction rules. Along the same lines, concurrent strategies of improvement and professionalism may be introduced to develop public agencies as well as flexible and diversified private participation schemes, such as the ones proposed by PROMAGUA programme. As Gérard Mestrallet, Chairman and CEO of Suez Company says in a brief article: "It is necessary to overcome the inadequate dogmatic debate in order to reach the conditions that will enable us to supply those 1200 million people who are currently needing water in the world" (Mestrallet, 2002). Nevertheless, it is also crucial to accept that there are many examples of public systems that have long-term planning and work with financial performance and stability levels far superior to those of many private companies. Therefore, the federal and state government authorities should not simply walk away from this priority field of service.

According to Surita Narain (quoted in Henderson, 1993), from The New Delhi Center for Science and the Environment in India, the key requirements for change in this industry are public participation and political processes demanding good governance on water management issues, instead of relying in new service contractors. In other words, if an efficient and co-operative administration with good budgetary results working under the correct legal framework cannot exist due to the prevailing circumstances, then fostering private investment agreements as a mechanism to counterattack the discretion and inefficiency of some government circles might be the wrong strategy, and sooner or later it will demand the allocation of public resources for rescuing those schemes lacking adequate support.

Developing a Strategy to Achieve Stability and Administrative Consolidation of the Water and Sanitation Services: Potential Roles of the Private Sector

Departing from the desirable conditions required to achieve financial and operating sustainability of the water systems, Table 4 presents a combination of actions that will constitute an improvement strategy leading to the consolidation and administrative soundness of the systems. This is a preliminary formulation that does not pretend to exhaust all the development options that may lead to a successful implementation of different schemes for capital and operating costs recovery, by stabilizing the technical and administrative capabilities as well as the required social values and institutional elements.

Table 4. Comprehensive strategy to strengthen the water and sanitation sector

	Current situation	Short term	Medium term	Conditions
1. Legal framework	Service concession and contracting basis are clear	The Water Act has to be amended to include regulatory and guarantee processes.	CEAG will be established as a regulating entity	Political consensus and the expression of needs
2. Technical risk	There are cadastral surveys and adequate plans. The law must provide for recovery at the rate's level.	The agencies planning efforts should be based on sound technical information.	Special financial mechanisms must be created to keep abreast with the infrastructure replacement works, without affecting the financial position of the system.	CEAG must continue providing technical support. It is necessary to establish a professional certification mechanism and to implement technological assistance programmes.
3. Public and political acceptance	There is no clear social awareness about the technical and financial implications of the service supply. There is no information regarding the system's situation nor about the financing alternatives.	To inform about the complexity or problems related to management, operation and financing of the service. Information about financing issues is required. Political bias has to be discouraged.	Public disclosure of the financial results and performance recognition must be held as positive socially accepted values. Lack of administrative continuity and inefficiency ought to be considered as negative values.	Obtain support from the mass media.
4. Income stability	The rate structures are varied and inadequate in many cases	To create a standard state structure already negotiated with the Congress. To establish 'safety locks' in the regulatory framework to ensure service sustainability	To enforce the technical expert's decisions on rates and the supplementary mechanisms used for presenting the agencies' financial results. Financial support mechanisms are required.	Political consensus. Broad communication with opinion leaders and the general public.

5. Investment optimization	CEAG implements a programming system based on indicators. Internal revenue investment does not follow uniform priority and optimization criteria.	To establish hydraulic programming systems in 20 cities with more than 20 000 people.	Integration of financial support mechanisms for hydraulic development, after a careful analysis of social profitability.	Transformation of the domestic financial system. Creation of a state financial system.
6. Leverage minimiza-tion	Few municipalities maintain onerous loans; they do not want any more liabilities. Operating leverage varies.	Feasible mechanisms have to be created to reinforce Long-term financing. Promoting adequate standards for fixed expenses.	To prepare inter-municipal mechanisms for risk management and risk sharing.	Political commitment of the three Powers in Office. Reliable training, evaluation and certification mechanisms.
7. Human capital improve-ment	There are isolated training efforts. CEAG promotes occupational certification.	To make a professional certification project. Expand the existing occupational certification project	Fostering the participation of experts in professional colleges. The job stability of certified staff will be a budgetary priority.	
8. Ensure future supply	There is a proposal for a hydraulic programme	To obtain pending allocation and reserve Certificates	To create a strategic bank of 'Reserve certificates'. Use the economic power of other sectors in public supply	To keep the basins in good operating condition. Political will and federal support are needed. Good weather conditions.
9. Co-ordina-tion mech-anisms	There are no proposals about the government entity that should rule over private investment. The regulatory and enforcement powers of the three government levels are full of ambiguity.	To amend the Water Act and publish regulations including regulatory and co-ordination concepts.	CEAG will act as a regulating entity. Determine clear rules for mediation and decision making issues.	Federal support. Political will of the parties.

The private sector can take part of this development at different stages, depending on the viability and convenience of this interaction.

Final Comments

The federal and state legal frameworks accept the state's intervention in water and sanitation services when, at the request of the municipal administration and with the Congress authorization, the municipality is considered incapable of performing those services. On the other hand, the state government, by means of its decentralized entities and agencies, is responsible for the harmonious development of the state. Appropriate supply of quality water in the amount and timing requested is a crucial social and economic development factor in Mexico and Guanajuato. Therefore, the state government takes an active part as a supportive and subsidiary support for the development of the operating agencies.

Within this framework, private investment is conceived as a promissory alternative to supply quality service with access to alternative sources of finance, provided the minimum administrative and environmental conditions exist. However, the implementation of this type of scheme must be regarded with caution, avoiding any risk transferring to the state government whenever the institutional development propitiates the situation, by means of service agreements or private investment mechanisms that are only considering particular risks without encouraging deep changes in the institutional structure of the sector. Active participation of the State Commissions as linking counterparts between the federal regulatory agency and the state or the municipality or a combination of both entities, is an additional ingredient for success that must be taken advantage of. On the other hand, the users' participation is also a condition for achieving water governance, including urban public users. All these conditions must be fulfilled for the efficient implementation of private participation as a convenient option and not as a necessary tool to transform this sector. Such an option must emerge from the municipal level, as a result of intensive dialogue and decision-making work carried out directly at the operational level of the systems.

References

Amatetti, E. (1994) Managing the financial condition of a utility, *AWWA Journal*, April, pp. 176–187.

Brook-Cowan P. (1997) The private sector in water and sanitation – how to get started, in: *The Private Sector in Infrastruture – Strategy, regulation and Risk*, pp. 89–92 (Washington, DC: The World Bank).

CNA, BANOBRAS, SEMARNAT (2001) Programa para la Modernización de Organismos Operadores de Agua (Programme for the Modernization of Water Utilities), internal document, National Water Commission.

CEAG-LARUZ (1997, 1998, 1999, 2000, 2001) *Diagnósticos sectoriales de Agua Potable y Saneamiento (State Water Sector Diagnosis Report)*, annual editions, Guanajuato State Government.

CEAG-LARUZ (2001) *Análisis y propuestas al proceso de los decretos de condonación de adeudos por pago de derechos de explotación y vertido (Analysis and proposals for the process related to the Presidential Acts for the exoneration of water use and discharge rights to the muncipalities)*, executive synthesis of the project report, internal document.

CEAG-LARUZ (2002) *Sistemia Tarifario Integral para el Estado de Guanajuato (Integrated Rate Structure for the State of Guanajuato)*, project report.

Guanajuato State Congress (2000) *Compendio de Leyes y Reglamentos del Estado de Guanajuato (Guanajuato State Laws and Regulations Compilation)*, LVII Legislature, Guanajuato, CD ROM version.

Henderson, B. (1993) Water privatization – success or failure?, *Pipes & Pipelines International*, January–February.

Idelovitch, E. & Ringskog, K. (1995) *Private Sector Participation in Water Supply and Sanitation in Latin America*, (Washington, DC: The World Bank).

INEGI (2000) *Censo General de Población y Vivienda (General Population and Housing Census)*, Government of Mexico.

Mestrallet, G. (2002) Clearing the air, *Water & Wastewater International*, 17(2).

Mould, M. C. (1980) Financing water resources development, *Water Resources*, pp.199–211.

Concepts of the Chilean Sanitation Legislation: Efficient Charges and Targeted Subsidies

DAMARIS ORPHANÓPOULOS
Ministry of Public Works, Santiago, Chile

Introduction

Towards the end of the 1970, some new ideas were considered in Chile in terms of how best to improve the management of the public utilities. The market economy and the related economic models appeared to have some application to improve the performances of these utilities.

In terms of water supply and sanitation services, there were many problems that needed to be solved in order to achieve some goals that seemed desirable from every point of view. Continuing deficits in the operation of the water utilities meant a subsidy had to be provided to all users on a continuing basis, and often for a bad quality and/or inefficient service. The universal subsidy meant that benefits were being provided to families that could afford to pay the real costs of the service. Socially, water and sanitation services were required to be of good quality, and to guarantee an adequate coverage of drinking water supply and sewage disposal for the entire population. On the other hand, there was no justification for keeping the universal subsidy on a continuing basis.

The new Chilean legislation covering the water utilities has met many of these constraints, creating an institutional framework that is able to comply with these new requirements. On the one hand, the legislation offers the companies that provide water and sanitation services the legal support for an appropriate profit on their investments. On the other hand, it offers all users the legal support for an appropriate service at reasonable costs.

One important aspect that needs to be considered is whether all the users are economically able to pay these new higher charges. In order to solve this situation, the relevant legislation has provided for a subsidy for drinking water supply and sewage disposal for the poorest users. Proper implementation of this law is the key to maintain and operate an efficient system.

Legal Framework

It is necessary to analyse the social, economic and political framework within which this new legislation for water utilities was developed.

The legislation responds directly to a clear political will of the government to improve health and quality of life of the population in a sustainable manner. It was also decided that the public utilities should be provided by an efficient private system instead of the public sector. The fact that the water utilities are natural monopolies should not prevent the application of market economy principles to improve their performances. The only condition was that the charges agreed to for water and sanitation services would allow the companies to obtain a reasonable economic return that would make their participation attractive. Accordingly, public utilities in general were sold to the private sector, like the communication and electricity sectors earlier.

While above considerations are important, it should be noted that the political will and the economic support do not fulfil all the necessary conditions to achieve all these goals. The political will must result in the formulation of a set of coherent and clear activities that can allow the whole system to work in an expected manner. In relation to the water services, what reflects the commitment of the government is the establishment of a clear and coherent legal framework that is able to contribute in the creation of a sustainable private sector in terms of:

- Legislation on water rights that regulate abstractions and use of waters.
- Water quality norms that regulate the quality of water for various uses, and the allowable qualities of effluents to be discharged.
- Health and environment protection legislation.
- Sanitation legislation.

The legal framework thus establishes the relationships and commitments between the private sector and the water users, and empowers the government to force each party to comply with the legislative requirements. With regard to drinking water supply, the most important consideration is to develop the awareness in the population that drinking water is a 'product', and that there is a cost to obtain it just like any other product, and that this cost should be paid by the consumers themselves based on their actual consumption. It is not possible to have sustainable services for water utilities within a context in which the population perceives that drinking water supply and sanitation should be free.

A framework for a three-party commitment was established. First is the company that assumes the responsibility of supplying drinking water supply and sewage disposal services on a long-term basis. This responsibility can only be achieved if the users assume the responsibility of paying the appropriate charges for the services provided by the companies, and on time. The third party is the state, which guarantees that each party complies with its obligations.

Laws and Concepts for Water Utilities

The legal frameworks for the operation of the water utilities in Chile include five fundamental laws (with further legal modifications), and the regulations for their implementation. These five laws are briefly discussed.

The Law that Establishes Charges for Drinking Water Supply and Sewage Disposal Services (DFL No. 70 of 30 December 1988)

This law considers two issues. First is the provision of drinking water supply and sewage disposal as services, which should generate an assured profit in order to ensure the long-term survival of the providers of these services. In addition, the drinking water supply and sewage disposal services are natural monopolies, which can result in very high charges unless the services are properly regulated. Hence, the law establishes that the charges for these services should be set in an appropriate manner, and it further stipulates a procedure to calculate these charges appropriately. This procedure is used by the Office for Water Services to determine the maximum price the companies that provide the services can charge.

Nature of the charges. The costs considered in the formulae to determine the charges are calculated on the basis of a 'model company'. A 'model company' is one which provides services in an efficient manner, and one which follows the normative framework and the regulations that are in force, as well as the geographical, demographic and technologic restrictions for its operation. In addition, it must consider the potential relationships among the different companies that provide the service.

The costs that are considered to estimate the charges are based on the costs for each of the water services that a 'model company' would have to incur. In addition, each 'model company' has to develop an institutional and administrative programme that includes the different roles of the company which provides these services. It should consider the integration of the several steps for providing the services, with the objective of minimizing the costs. The 'model company' also has to develop a physical plan for the system for all the phases of the water services provided.

The investment costs must be separated for each component of the water services provided. For example, the cost of raw water is determined through water rights market, and is considered to be an investment when calculating the incremental cost of development as well as the long-term total cost. If the Office for Water Services decides that the market values of the water rights do not reflect the opportunity costs of that resource, or, if the existing information is insufficient, the cost of the raw water will be calculated based on its implicit value inherent in the price of the land.

The design of the facilities must explicitly consider the economically-efficient losses and the security levels required to provide the services.

The formulae to determine the water and sanitation charges include the costs and their indexation mechanisms, and are estimated based upon the incremental costs of development. The incremental cost of development is defined as the value equivalent to a constant unit price, which, when applied to the future incremental demand, generates the necessary income to cover the incremental costs of efficient exploitation, and of investment requirements for an optimized expansion project. In other words, the net present value of the expansion project is equal to zero. The expansion project should be carried out in not less than 15 years. This calculation considers the economic life of expansion-related assets, appropriate rates for taxes, and a capital cost rate that corresponds to the long-term average rate for the national currency that is offered by the Central Bank, plus a risk factor of 3 to 3.5%. The capital cost rate cannot be lower than 7%.

If there are no plans to expand, the formulae to determine the charges are determined on the basis of the long-term marginal costs. The long-term marginal cost of a service is the incremental long-term total cost of providing it, considering the increase of one unit of the provided quantity. The long-term total cost is the constant annual value required to cover the efficient costs of exploitation and investment of an optimized replacement project for the company, dimensioned to satisfy the demand, and which results in a net present value equal to zero within a time horizon of not less than 35 years. The methodologies for estimating these costs are provided in detail in the regulations of the law.

The pricing process.   The formulae that define the charges for each entity have a validity period of five years. The two exceptions to this time limit are when there is an agreement between the entity and the Office for Water Services, to extend the period for another five years, using the same formulae. It is based on the assumption that no changes are needed for the calculations. This agreement must be done 14 to 17 months before the five-year period expires. The second exception is when there are important modifications in the assumptions for the formulae, which justify the initiation of the next pricing cycle.

The estimation process for pricing should be initiated 12 months before the end of the period in question. At that time, the Office for Water Services must publish the Terms of Reference for the study that will decide on the formulae for the next period, for the information of both the companies and the general public. Within the 45 days after this information is published, the Office must answer all the questions and comments received.

The studies that the Office carries out must be based on the behaviour of the 'model company'. This is efficient management with optimized expansion plans. This means a guarantee for the consumers that only the minimum necessary costs would be considered for the provision of the services.

During the pricing estimation process, the Office establishes charges based on this law, and uses the methodologies established under the regulations. These charges are appropriate since they are based on the long-term marginal costs. The income that these charges may generate, considering the present annual demand, must then be compared with the long-term total cost of the business. Generally, the income that is generated with the charges does not cover the long-term total costs, and thus has to be adjusted to cover them. The estimated charges then have to be readjusted by discounting the replacement values of the facilities, amounts contributed by third parties, again considering their replacement value, and then transformed into annual values. In this way, the final prices are obtained. Finally, these tariffs are expressed in formulae that incorporate the representative price indexes of the cost structures involved in the different phases for the provision of water and sanitation services.

Simultaneously the companies are expected to undertake their own pricing studies using the same considerations as the Office. During this price determination process, the studies of both parties are reviewed by each other. If there are no discrepancies, the charges estimated by the Office are adopted. If there are differences, the company must make a formal presentation of these differences to the Office. If the discrepancies cannot be solved directly with the Office within 45 days of the exchange of the studies, the Office must formulate a commission consisting of three experts, who are to be paid by both the Office and the company. This commission must analyse the discrepancies, and then take a decision on the two estimates. The decision of the commission is final for both parties. Once the pricing process is completed, all the studies, calculation procedures and reports must be published to assure the transparency of the process.

The charge formulated must be approved by a Supreme Decree of the Ministry of Economy. This tariff structure will last for a period of five years, during which the charges are recalculated automatically every year, using the appropriate indexes stipulated by the law.

Calculation of the charges. The charges established by this law are based on marginal costs and calculated separately for each step of the water services, which includes abstraction, production, and distribution of water and disposal of wastewater. The objectives are to optimize the use of the resources, and to consider the possible interaction of services in order to minimize the long-term costs of providing the service. The estimation process for the charges has the following five steps.

1. Marginal costs. First the individual marginal costs for each step of the service are estimated. For each step, the following costs must be determined: the periodical fixed costs, variable costs associated for volume production during non-peak periods, variable costs associated for volume production during peak periods, and the costs associated with the capacity of the company, which represent the investments necessary for the expansion. The non-peak period is represented by eight months of low consumption. This could be six months for some regions. The remaining months of the year represent the period of high consumption or the peak period. The consumption during the non-peak period generates the normal average value. Consumption above this value is considered to be an over-consumption during the peak period. The difference between consumption and over-consumption is important, because the costs of the different concepts apply to each one of them. Consumption under $40\,\text{m}^3$ is not considered to be an over-consumption, even when it goes above the average consumption value. The marginal costs finally considered are those detailed in Table 1 (by phase and by m^3). At present, for the abstraction and disposal services, the companies work with unique values for peak and non-peak periods.

2. Efficient tariffs. Later, these marginal costs are transformed into efficient tariffs in the light of the following formulae contained in the regulations of the law.

The variable tariff associated to the volume of each one of the phases in the non-peak period, CVinp, is obtained as the corresponding variable cost, plus a part of the cost of capacity:

$$CVinp = CVi1 + Np/12*CVi3$$

The variable tariff associated to the volume of each one of the phases, in the peak period, CVip, is obtained as the corresponding variable cost, plus a part of the cost of

Table 1. Individual marginal costs for water services, per phase

Phase of the service	Fixed cost $/m³	Volume assoc. costs non-peak	Volume assoc. costs, peak	Capacity assoc. costs $/m³	Average costs $/m³
Drinking water production	—	CVP1 O&M costs of prod. Vol. non-peak	CVP2 O&M costs of prod. Vol. peak	CVP3 Investment in production development plan	—
Drinking water distribution	CFP = Admin. expenses per household; CFC = Admin. expenses per consumers	CVD 1 O&M costs of distr. Vol. non-peak	CVD2 O&M costs of dist. Vol. peak	CVD3 Investment in distribution development plan	—
Sewage collection	CFR = O&M expenses per household	—	—	—	CVR O&M costs, total volume
Sewage treatment and disposal	—	—	—	—	CVT O&M costs, total volume

Table 2. Efficient tariffs obtained from marginal costs

Phase of the service 1	Fixed charges $/m³	Volume charges non-peak $/m³	Volume charges peak $/m³	Over consumption charges $/m³	Average charges $/m³
Production	CFP	CVPnp	CVPp	CVPsc	
Distribution	CFC	CVDnp	CVDp	CVDsc	
Sewage collection	CFR				CVR total cons.
Sewage treatment and disposal					CVT total cons.

capacity:

$$CVip = CVi2 + Np/12*CVi3$$

The variable tariff associated to the over-consumption volume, CVOC, is obtained as:

$$CVOC = CVi2 + CVi3$$

In the formulae, the letter i represents the phase of the service, and Np is the number of peak months in the year. Fixed tariffs are the fixed costs, and no transformation formula is applied to them. In this way, the unit costs matrix is transformed into a unit tariff matrix, which is shown in Table 2.

3. Adjustment and correction of efficient charges: final charges. Subsequently, efficient tariffs must be corrected as noted earlier in order to cover the long-term total costs of the business, if the total income generated by applying these charges for the existing annual demand do not cover them. Besides, these tariffs must be adjusted by the contributions from the third parties. These adjustments and corrections result in the establishment of the final tariffs. The final tariffs show the same structure as the efficient charges of Table 2, and are shown in Table 3.

4. Indexing of final charge. For each charge, the law provides a price index, which allows automatic adjustments during the five years of operation of the formulae. These price indices, or indexation factors, apply directly to each one of the final tariffs in the matrix of Table 3. These are shown in Table 4.

5. Official charges. The information in the matrix for the indexed final tariff of Table 4 is the content of the Supreme Decree that approves the water tariff for each company at the end of a pricing process. These charges must be published in the newspapers each year, and then collected from the users on a monthly basis. The final tariffs are grouped in the following manner, producing the official charges that are expressed in terms of five parameters:

- Fixed charges. It is the addition of the indexed final fixed charges of Table 4 ([2] + [3]). There are no fixed charges for the phases of production [1] non-disposal [4]. The fixed charges are commercialization costs, and operation and maintenance costs for drinking water and sewage collection from the households. These costs do not depend on the volume of water. The fixed charge is the same for all the consumers of the same company.
- The variable fee for the consumption of drinking water during the non-peak period includes the volume associated with the costs of production during the low consumption period ([5] + [6]). It is applied to the actual quantity of water consumed during the non-peak months.
- The variable fee for consumption of drinking water during the peak periods covers the volume associated with the costs of production and distribution during the high consumption period ([9] + [10] of Table 4). It is applied to the actual quantities of water consumed during the summer, which do not exceed the average consumption of the winter or a maximum of 40 m^3 if the average consumption of winter is smaller.

Table 3. Final tariffs: adjusted and corrected efficient tariffs

Phase of the service	Fixed charges $/m³	Volume charges non-peak $/m³	Volume charges peak $/m³	Over-consumption charges $/m³	Average charges $/m³
Production	CFP	CVPnp	CVPp	CVPsc	
Distribution	CFC	CVDnp	CVDp	CVDsc	
Sewage collection	CFR				CVR total cons.
Sewage treatment and disposal					CVT total cons.

Table 4. Indexed final tariffs and official charges

Phase of the service	Fixed costs $/m³	Volume fares non-peak tariffs $/m³	Volume tariffs peak $/m³	Over-consumption charges $/m³	Average tariffs $/m³
Production	[1]	[5] $CVPnp \times INP1$	[9] $CVPp \times INP2$	[13] $CVPsc \times INP3$	[17]
Distribution	[2] $CFP \times INFD1$ $CFC \times INFD2$	[6] $CVDnp \times IND1$	[10] $CVDp \times IND2$	[14] $CVDsc \times IND3$	[18]
Sewage collection	[3] $CFR \times INFR$	[7]	[11]	[15]	[19] $CVR\ tc \times INR$
Sewage treatment and disposal	[4]	[8]	[12]	[16]	[20] $CVT\ tc \times INT$
Formulae to calculate the tariffs	$A = \overline{2 + 3}$	$B = 5 + 6$	$C = 9 + 10$	$D = 13 + 14$	$E = 19 + 20$

- The variable charge for over-consumption of drinking water during the peak period covers the costs of capacity investment for the phases of drinking water production and distribution ([13] + [14] of Table 4). This fee is applied to over-consumption during the summer period. Over-consumption is considered to be those cubic metres that exceed the average winter consumption, or a maximum of $40\,m^3$, if the average consumption for winter is smaller.
- The variable fee for sewage collection, treatment and disposal services is based on the same quantity of water that is consumed in each household. This charge covers the additional costs of the phases of sewage collection and disposal ([19] + [20]). At present this charge is the same for both normal and over-consumption of water.

Table 5 shows the actual charges for the 16 companies for the most important Chilean cities.

Reimbursable financing contributions. The service companies may require reimbursable financial contributions from new clients requiring new connections, or from those clients who request an increase in water supply, for which the company must increase its capacity. The financial contributions for increasing the capacity may be collected provided that a new service, or an enlargement of an existing one, is not included in the development plan of the company. The financial contributions can be collected to cover extensions of the network beyond the existing facilities, including extensions that are considered within the development plans. In this case, the extension must be designed so that it is technically compatible with the development plans.

Financing contributions can be reimbursed to the consumers in any way that is mutually agreeable. These could be reimbursed in cash, commercial papers, shares of the company, provision of drinking water and sewage services, or by means of any other mechanism the two parties can agree upon. The form and term of the reimbursement are stipulated in a contract that is signed by both the parties.

In summary, the structuring of water tariffs are based on the following major considerations:

- The charges are structured in such a way that they should cover the costs of an efficient and effective service-provider working 100% of the time, without deficit and with the minimum possible cost for the users.
- The minimum cost for the user is guaranteed through the consideration of a parallel pricing process, conducted simultaneously by the company and the Office for Water Services. Charges must be based on the operation of a 'model company'. The company cannot transfer the costs to the users because of inefficient management, non-optimized expansion plans and other avoidable costs, even if it is possible to transfer economically efficient losses.
- The implementation of the system is legally binding. Failures can result in the termination of the concession. The water requirements of the users should be carefully assessed, and the designs made accordingly so that the demands can be satisfied under all conditions.
- No deficit in the operation of the company is achieved through the adequate consideration of all costs: planning, construction, replacement, operation, maintenance and expansion of the system. The estimation of the charges is done

Table 5. Charges levied by main water companies in Chile, per region, Ch$/m³

Charge	A	B	C	D	E
Company	Fixed charges	Charge per m³ non-peak drinking water	Charge per m³ peak drinking water	Over-consumption peak drinking water	Charge per m³ Sewage
ESSAT SA I Region	570	505	548	1070	188
ESSAN SA II Region	728	860	856	2152	194
EMSSAT SA III Region	668	259	238	464	245
ESSCO SA IV Region	605	281	212–355	549–1006	no treat. 181–305 with treat. 239–416
ESVAL SA V region	613	421	408	789	240
A. Andina Metropolitan Region	442	201	194	495	117
Aguas Cord. SA Metropolitan Region	552	239	230	412	107
Maipú Metropolitan Region	506	153	150	376	143
ESSEL SA VI Region	450	234	232	550	294–343
ESSAM SA VII region	530	212	206	434	244
ESSBIO SA VIII region	605	233	227	477	228
ESSAR SA IX region	575	239	233	577	205–233
Aguas Décima SA X region	354	217	215	550	381
ESSAL SA X Region	436	300	—	—	278
EMSSA SA XI Region	811	517	—	—	547
ESMAG SA XII Region	770	410	—	—	237

in such a way that the company can operate without deficits. This includes determination of an efficient tariff so that the average long-term costs can be covered. This fact guarantees the most desirable characteristic of a drinking water supply and sewage disposal system: the long-term sustainability of the system.

- The companies must operate without profit over the long-term. They can operate with an enhanced capital rate of return, in relation to the prevailing base rate of the market.
- The cost analysis for each phase of the service is transparent.

Law of Subsidies for Water Consumption and Sewage Disposal (Law No. 18.778 of 2 February 1989 and its Regulations)

This law establishes two types of subsidies: a consumption subsidy and an investment subsidy, which benefits primarily rural areas. The first subsidy is financed directly through the national budget, and the second one is financed through the annual budget of the Ministry of Public Works.

Consumption subsidy. Initially the state subsidy for consumption of drinking water and sewage disposal targeted the 20% of the poorest families of each region. In order to determine who were the poor people and where these people lived, a National Socio-economic Characterization Survey (CASEN) was carried out for the first time in 1987, and the results of the historic files of the Unique Family Subsidy beneficiaries were used. The subsidy law establishes the way in which the state will pay a part of the service bills of the beneficiaries. This payment is made by the municipalities directly to the water company of the region. The subsidy law was created in conjunction with the law establishing the tariffs, which assures the company that the charges can be collected to assure their long-term survival. This does not consider the poverty level of the users.

The law establishes a subsidy for the poor to cover the charges for drinking water and sewage disposal services. This subsidy is also available to users of drinking water services only. The subsidy applies to both fixed and variable charges. Although initially there was a consumption restriction of $20\,m^3$/month, at present this limit is no longer a condition for applying for a subsidy. The subsidized percentage of fixed and variable charges varies between 25% and 85%, and should be the same for the beneficiaries of the same region who are charged the same charges and belong to a similar socio-economic level. This subsidy can be used with other municipal subsidies.

Conditions to obtain and to maintain consumption subsidy. The subsidy must be requested by filling in a form in the municipality. The applicant must be up-to-date in the payment of the water service bills. The applicant must belong to the group of people who cannot afford to pay for the services. This is established by means of the CASEN Survey on the average income of the population, information obtained from the various institutions given pensions, and with the criteria of the Pan-American Health Organization that recommends that not more than a 5% of the average monthly income of a family group should be used to pay for drinking water services at the national level.

In addition, the Planning Ministry (MIDEPLAN) has a Social Information Department with regional offices, which assigns a score to all potential beneficiaries of this subsidy, in order that the subsidies can be targeted to the poorest. The subsidy does not apply under the following conditions:

- A resident who moves to another municipality.
- A resident who moves within the same area, but without notifying the municipality.
- Delay in paying the non-subsidized part of the bill.
- Non-fulfilment of any of the conditions required for subsidy.
- Expiry of a three year-period, after which it is necessary to apply again.
- Voluntary resignation by the beneficiary.

Evolution of law of consumption subsidy. The law on consumption subsidy is the conceptual compliment of the law covering the tariffs, which considers charges that should cover the real costs. When part of the population cannot pay these charges, and the bills remain unpaid, the water companies cannot survive under these economic conditions.

The subsidy law was initially conceived as being very strongly focused, whose implementation required an exhaustive analysis of the socio-economic profile of the entire population, and of the poverty conditions of the beneficiaries, in order to target these subsidies to the needy. Furthermore, the law also required that each beneficiary who complied with the conditions must be an applicant, and also to take the personal initiative to request it.

The law has been in operation since 1990. During the first years of operation its impacts were limited due to lack of knowledge of the people of the legal requirements, and also the lack of a culture of requesting subsidies. Nevertheless, since 1993, the effective disbursement of this subsidy has approached what was initially estimated (over 90%). To improve the effectiveness of this subsidy, the government subsequently modified the law several times by relaxing the conditions to obtain it. The later legal modifications eliminated the consumption limit of $20\,m^3$/month, increased the amount of water consumed that could be subsidized, made more flexible in terms of the percentages of people who could be subsidized, made it possible for secondary dwellings where the main household is not eligible for the subsidy, and granted the water companies the authority to apply for the subsidy in the name of the users who cannot comply with the conditions to obtain it. In this way, the subsidy was made available to a larger number of economically disadvantaged people than before.

At present, the subsidy is granted to the level about 95% of the original estimate. Even though it is not as targeted as it was initially planned, it still is a focused subsidy, and, for the most part, it is being efficiently allocated. This subsidy ensures that the income that could be obtained from the poor households via tariffs represents real incomes for the companies. This is an essential condition so that the companies can cover their cost of financing the services provided, and also operate in an efficient manner.

This subsidy is a better and fairer way for the state to allocate its scarce resources directly to the poor who need it, rather than continuing with an universally subsidized system, irrespective of the economic needs of the people.

Some of the practical difficulties in managing the implementation of the subsidy programme include transfer of appropriate funds from the Ministry of Finance to the municipalities, and from the municipalities to the water companies, in a timely manner. This is because the Ministry requires that the funding expenditures be justified beforehand.

In addition, the low capacities of the municipalities to manage administrative issues, makes them somewhat inefficient to apply properly the legal restrictions for the subsidy for the different conditions.

Subsidy for investment for rural drinking water systems.   Another modality of the so-called subsidy law was established for rural water systems. This is known as the investment subsidy. The objective of the investment subsidy is to carry out pre-investment studies in order to improve the existing rural water systems, and to cover the difference between the total cost of investment and the contributions of the interested users (and other agencies) according to their capacities to pay.

This subsidy theoretically works in the following way. Annually, the Ministry of Public Works makes a list of a programme for investments based on actual costs for providing rural water and sanitation services, or for improvement of the services to a national level. This programme is based on the studies of the Ministry, and by the needs presented by the communities.

In addition, the Ministry of Public Works carries out an evaluation of the socio-economic conditions and payment capability of each interested community. This must be carried out according to a methodology that must be legal and in agreement with the Ministry for Planning (MIDEPLAN) requirements. As a result of the investment calculations and consideration of the payment of the interested communities, this Ministry publishes a list informing each community of the total investment that is needed, payment capacities, and the amount that could be being granted.

The communities that decide to apply for the subsidy must make a formal request to the Ministry, in a pre-established format. These requests are then evaluated by the Regional Department, according to the following criteria of eligibility: high socio-economic benefits from the schemes as measured by traditional project evaluation methodologies; existences of low socio-economic conditions in the project areas; and payment capacities of the applying communities.

The projects that comply with these criteria are selected and presented to the National System of Investments for the Public Sector, and to the Ministry of Finance. The resources allocated for the payment of the subsidies for these projects come from the budget of the Ministry of Public Works, and are allocated at a regional level. The regional governments must prioritize the projects for the regions concerned, and decide on the distribution of the financial resources among the projects in the rural areas that comply with the criteria of eligibility indicated above. However, the problem is that the investment subsidy is not yet operational. Although the regulations for the law exist, as well as the socio-economic community evaluation methodology, the rural water systems are being developed directly with a fiscal contribution that the national budget allocates to the Ministry of Public Works for these specific purposes. This fiscal contribution does not pass through any targeted process, as was intended by the law.

General Law of Sanitation Services (DFL No. 382 of 21 June 1989)

This law includes the following considerations.

Concession Regime

The law establishes that all the public or the private companies that provide water and sanitation services will be under the concession regime. Only the smallest services, with less than 500 household connections will be excluded from certain norms of fiscal control, as well as from some restrictions that are applicable to other monopoly-type public utilities.

The concessions allow the installation, construction and operation of water services for the production and distribution of drinking water, and the collection, treatment and disposal of sewage. The concession is awarded either on request or through bidding by a society that must be legally formed according to the rules of the country, and with the only objective of providing water and sanitation services for which it was created. In the request for the concession, all characteristics of the requested concession must be provided, as well as a guarantee that varies between 50 and 1500 UF depending on the number of connections.

The concession request must include at least the following information:

- Type of concession.
- Identification of water sources, and water rights that the company can use (these rights must be property rights or rights to use water).
- Companies to which it will be related.
- Geographical area that will be covered by the company (this area may be increased by the Water Services Office).

In addition, if there are two or more one companies requesting concessions, all of them must present the following information for evaluation by the Office:

- Development plan (schedule of the works and calculation of benefits, costs, net value and associated profit) to cover the requested services, based on a techno-economic pre-feasibility study.
- Proposed charges and expected reimbursable financing contributions.

The decree for the concession contains all the characteristics and technical conditions that the company must comply with, as well as the charges, the norms that the company will have to follow, and the guarantees involved. This decree then becomes a legal public document. The water distribution companies are responsible for the collection of charges, and for other services such as abstraction and production of water, and sewage disposal. At first the concessions have no time limit, but they expire if the companies do not comply with the law or with the concession decree.

Rights and obligations of the companies and users. The company is obliged to provide the service within the area defined in the concession. At first the company can be required to expand its services and to modify its development plans by the Office for Water Services, provided this Office has reasons to demand such changes. The company must maintain water quality at its own expense. In addition, the company must guarantee the continuity and the quality of the service, unless some unforeseen situations arise.

The rights of the company include the following:

- Collect the charges from the consumers for the services, and required reimbursable financing contributions.

- Collect current interest rates for delays in the payment for the services received.
- Suspend the services to consumers, subject to previous notices if debts are not paid, and to collect the charges for suspension and reinstallation.
- Collect the charges for repairing damages caused by the users.

The users of the sewage networks are not allowed to discharge substances that could be harmful to the system, or interfere with the treatment processes or that are prohibited by the norms. The company has to right to cut the services to those users responsible for any of the previous infringements, as well as charge for monitoring the fulfilment of the appropriate regulations.

Monitoring and control. The companies that provide water and sanitation-related services are monitored by the Office for Water Services. This Office has the authority to ask for reports, inspect the services, require the development plans for projects, audit the accountings, and to implement all necessary measures in order to assure the fulfilment of the legal requirements and regulations in force.

According to this law, the concession system allows the inflow of private capital for the development of the water and the sanitation services. At present, several private companies have obtained concessions to provide water services to several cities of the country. This law assures that users have the rights to receive a good quality service at a competitive price, and to the companies the right to collect the appropriate charges. It bestows on the Water Office the fiscal authority to ensure that both the users and the companies comply with their responsibilities.

Laws Authorizing the State to Undertake Enterprise Activities for Drinking Water and Sewage Disposal and Define the Structure of Corporate Companies (Law No. 18.777 of 8 February 1989, and Law No.18.885 of 12 January 1990)

These two laws mark the beginning of the transformation of the water and the sanitation sector by allowing the state to undertake enterprise activities, and transforming the regional water utilities into corporate companies. Initially, with the first law, only two companies were created in this way for the regions V and Metropolitan. The second law refers to all other water services for the 11 remaining regions, which were transformed into corporate companies, and legally continued their regional water services.

These laws stipulate that the state, along with the Corporation of Production and Promotion (CORFO), initially would own 100% of the shares of these companies. Together, they could not own less than 35%, unless the Law of Societies would require otherwise. It is important to note that with a percentage higher than 10%, both the state and CORFO will always have to take the most important decisions jointly for the companies and important issues in the future (increase in capital, selling of assets, and other issues according to the law of corporations and the regulations of the company).

These laws establish that the goods and the real estate (movable or immovable) belonging to the public companies would become part of the new companies. The laws also require that the personnel of the public companies would become part of the new companies, but subject to the norms that apply to the private sector workers.

These laws introduce major changes in the system that have been in force since the 1980s. Initially, all the water utilities are transformed into corporations. In addition, not only the concept of an economic management and administration of the water services is introduced, (concept of efficiency), but also the concept of privatization (the state can sell whenever it considers appropriate). This way, the former central service is separated automatically from the companies which provide the service.

With regard to privatization of these services, the most important privatization in recent years was EMOS SA, where the state retained 49% of the shares. The state also retained the veto-right upon certain key decisions, such as the transfer of the water concessions and of the water rights that were part of the assets of EMOS SA, when the privatization operation took place. Certain strategic assets such as Laguna Negra, Yeso reservoir and Ramón valley were transferred to CORFO, before the privatization occurred. The company EMOS SA keep the exploitation rights of all the existing productive facilities. Finally, according to the Law 19.549, CORFO offered a 10% of the share capital to the workers of the company. Some 345 workers of EMOS accepted the offer, and acquired 3.4% of the share capital of the company.

The Law which Creates the Office for Water Services (Law No. 18.902 of 27 January 1990)

The Office for Water Services has the legal obligations for the National Service for Water Works in terms of regulatory functions, control of water and sanitation services and control of disposal of liquid industrial effluents. This way, the body providing services and the one implementing regulations, are separated.

The Office for Water Services was created as a legal entity, with decentralized roles and a budget of its own, under the surveillance of the President, but through the Ministry of Public Works. This Office has the authority to control, from the fiscal viewpoint, those companies that provide the services, implementation of the norms and disposal of liquid industrial effluents. To develop this control, this Water Office has created a series of norms (around 220), complementary to the basic norms of quality which the companies must comply with. The President of this Office must verify that the fiscal obligations are implemented and followed in the entire country.

The Office owns the resources that the law of budget of the Public Sector, or any other laws, authorize it, estimates annually the goods and real estate which originally belonged to the National Directorate for Water Services, or were acquired afterwards. The Office is authorized to impose fines on those companies that provide the services, for its own fiscal benefit. The fines could be imposed for the following reasons: (a) low quality of the service, use of unauthorized or undue charges, discriminatory-economic deals with users, inappropriate attention to the complaints received, damage to networks, or non-fulfilment of the obligation to deliver information to the Office according to the law; (b) infractions that put in danger or seriously affect the health of the population, or of the users in general; (c) infractions related to delivering false or clearly wrong information, and non-fulfilment of the law of concessions; (d) non-fulfilment of the development plans; and (e) release, or use unduly, the classified information to which the companies are entitled to.

The Office has the right to fine industrial or mining establishments that infringe the law, regulations, or norms related to the disposal of liquid effluents, or the non-fulfilment of instructions, orders and resolutions established by the Office. It even has the right to close any premises that do not comply with the discharge processing systems established by a Supreme Decree in a timely manner, or that infringe the law by discharging effluents near drinking water intakes, and hence putting in danger or seriously affecting the health of the population. The Office must also implement all measures that guarantee the safety of the population and which assures the rights of the users. It is also empowered to demand the support of the police to implement its decisions (infractions can be considered for up to four years after they are committed).

The Office must use all of the information provided by the different companies to decide the tariffs, mainly all studies and analyses, reports by several experts, updated development plans and infrastructural development plans as well as all other information that could be of interest to the urban planners and the users. The same obligation applies to the reports that the Office must issue regarding the quality of the services provided by the different companies, and any information that may be useful to the users. Finally, the Office has the obligation to keep an updated technical database for each company having a concession, including the basic parameters needed to calculate the real and optimized costs of each system.

Conclusions

The paper objectively analyses the advantages and disadvantages of the system that is now operational in Chile. The advantages for the users include efficient services, the right to complain, a transparent tariff system, consumers are not paying more than what is required, and access to appropriate subsidies. Among the disadvantages, there do not appear to be any for the users. So far, the attitude of the users has often been one of annoyance because they have had to pay the real costs for a good service. While the users perceive that the prices appear to go 'up and up', they are often not aware that services must be improved and expanded for the general benefit of the population. Initially, only drinking water was produced and distributed; later on, services for sewage disposal services had to be worked out, and finally, water treatment was made a reality. The system does not have any real disadvantage for the users, other than the fact that it is necessary to pay the real costs for a good service, which in the long-term may not be a disadvantage at all.

The companies have the advantage they can operate under a very clear legal and enforceable framework, assurance that the real costs can be covered, and that even if they do not generate any profit, they have an attractive rate of return on the capital. As in the case of the users, there are no disadvantages for the companies. A regulated profit is not a disadvantage. The state must regulate a public service that is a monopoly by nature, irrespective of whether it is public or private.

References and Bibliography

Chilean Construction Chamber (1993) *New Legislation on Sanitation* (Santiago, Chile).
ECLAC (1983) *Drinking Water and Environmental Improvement in Latin America, 1981–1990*. A work on some economic and social aspects (Santiago: Economic Commission for Latin America and the Caribbean).
Ministry of Economy (1990) Law that creates the Sanitation Services Office. Law No. 18.902 of the Ministry of Economy, Promotion and Reconstruction, 27 January, Santiago.
Ministry of Economy (2002a) Law that authorizes the state to undertake enterprise activities in matters of drinking water and sewage abstraction, and arranges the constitution of corporative companies for such effect. Law No. 18.777 of the Ministry of Economy, Promotion and Reconstruction, and subsequent modifications up to 2002, Santiago.

Ministry of Economy (2002b) Law that authorizes the state to undertake enterprise activities in matters of drinking water and sewage abstraction, and arranges the constitution of corporative companies for such effect. Law No. 18.885 of the Ministry of Economy, Promotion and Reconstruction, and subsequent modifications up to 2002, Santiago.

Ministry of Financing (1989) Water consumption and sewage abstraction subsidy law, law No. 18.778 of the Ministry of Financing, 2 February, Regulation of the law and subsequent modifications of both up to 2002, Santiago.

Ministry of Planning, Department of Planning and Cooperation, Social Division, Social Department of Information (1999) Administrative, operating and legal aspects of the water consumption and sewage abstraction subsidy, May, Santiago.

Ministry of Public Works (1988) Law that establishes the charges of drinking water and sewage abstraction services, DFL No. 70 of the Ministry of Public Works, 30 December. Regulation of the law and subsequent modifications of both up to 2002, Santiago.

Ministry of Public Works (1989) General law upon sanitation services. DFL No382 of the Ministry of Public Works, June 21, 1989, Regulation of the law, and subsequent modifications of both up to 2002, Santiago.

Mujica, R. & Lavanderos, M. (1981) Economic analysis of drinking water pricing by means of a simulation model. Document No. 72, September (Santiago: Catholic University of Chile, Institute of Economy).

World Health Organization (1995) *Financial Management of Drinking Water Provision and Sewage Abstraction Services. A Guide* (Geneva: WHO).

Zapata, J. A. (1981) Methodological aspects for drinking water pricing according to marginal cost criteria. Document No. 75, October (Santiago: Catholic University of Chile, Institute of Economy).

INDEX

AA *see* Aguas Argentinas
accountability 18, 19, 22–3, 26, 67, 79;
   Brazil 133, 139; monopolies 94; public-
   private partnerships 131; state-society
   synergy 137, 143
ad hoc models 98
administration, operation and maintenance
   (A, O & M) 19, 22, 26, 87, 168; *see also*
   maintenance
AGBA 57
Agencia Nacional de Aguas (Brazil) 6
Agenda 21 1, 7
agriculture 19, 21–2, 43; Brazil 104,
   123–5, 134, 135, 138, 139; Mexico
   217–18; Spain 33, 36; United
   States 41–2, 44; *see also*
   irrigation
Aguas Argentinas (AA) 60, 61–2, 71,
   73–4, 148–9, 154–8, 160–1
Aguas de Barcelona 57, 58, 59, 70
Aguas del Tunari 57, 58, 64, 72
American Water Works Association
   (AWWA) 45, 46
American Water Works Company, Inc. 88
Anglian Water 55, 56, 70
animals/livestock 9, 119, 129, 130
'appropriations doctrine' 43, 46
de Araujo, J. C. 104
Archibald, S. 47, 48
Argentina 5, 58, 77, 147–62; corporate
   compensation claim 63; currency risk 62,
   72; economic crisis 160–1; investments
   61, 62; regulation 65; renegotiation 60,
   65, 71, 72, 73–4, 156–7, 160; single-bid
   concessions 57; tariffs 70, 71, 72,
   154–60, 161

assets 46, 87, 91, 93, 153, 161n8, 185
associative networks 136
average referential price 13
AVSA 57–8
AWWA *see* American Water Works
   Association
Azevedo, Luiz Gabriel T. de 17–27
Azurix 55–6

Bahia 105, 107
Bakker, K. 65
Baltar, Alexandre M. 17–27
Belize 60
benchmarking 91, 95
Bhatia, R. 11–13
Biswas, Asit K. 1–6
block rates 13, 43; Canada 48; city users 50;
   United States 42, 45–6, 50
Bolivia 57, 58, 63, 64, 72, 77
Bonn Declaration (2001) 7
Braadbaart, O. 54
Braga, Benedito P. F. 117–30
Brazil 9, 14, 60; economic efficiency 20–1;
   price setting 97–115; public
   participation 131–45; size and diversity
   4; water charges 117–30; 'water
   poverty' 76
Brook-Cowen, P. 188
Budapest Sewerage Company 64, 67
Buenos Aires 5, 57, 60, 62, 63, 71, 147–62
bulk water provision: Brazil 104, 118, 119,
   121–3, 128, 130; guarantees 59;
   investment costs 26; marginal cost of
   supply 100; price setting 109, 110, 111;
   United States 43–4

Printed and bound by CPI Group (UK) Ltd, Croydon, CR0 4YY

01/11/2024

01782610-0005